Comeback Farms

Rejuvenating soils, pastures and profits
with livestock grazing management

Comeback Farms

Rejuvenating soils, pastures and profits with livestock grazing management

by
Greg Judy

A division of Mississippi Valley Publishing Corp.
Ridgeland, Mississippi

Library of Congress Cataloging-in-Publication Data

Judy, Greg, 1959-
  Comeback farms : rejuvenating soils, pastures and profits with livestock grazing management / Greg Judy. -- 1, ed.
      p. cm.
  Includes index.
  ISBN 0-9721597-3-8 (alk. paper)
  1. Grazing--Management. 2. Pastures--Management. 3. Livestock. I. Title.
  SF85.J83 2008
  633.2'02--dc22
                        2008013729

Cover designed by Steve Erickson, Madison, MS
Manufactured in the United States of America
This book is printed on recycled paper.

I would like to dedicate this book to Ian Mitchell-Innes of South Africa. In 2006, I was fortunate enough to hear Ian speak about Holistic High Density Planned Grazing used on his 14000 acre ranch in Africa. Ian really woke me up to the additional opportunities we all have on our farms if we will learn this method of grazing management.

Table of Contents

Dedication.....5

Acknowledgments.....8

Foreword - Painting pictures with livestock.....9

THE BASICS

Chapter 1: Education and Attitude - Starting with a Blank Canvas.....19

Chapter 2: The Toolbox - The Basics of Livestock Grazing Management.....24

Chapter 3: Where's the Grass? Uncovering the Farm's Hidden Potential.....31

Chapter 4: Our Free Fertility Program - Using Hay to Add Nutrients.....35

Chapter 5: Resist Opening Gates During Drought.....44

Chapter 6: Controlled? Burning.....49

Chapter 7: Why Switch from Stockers to Dry Cows.....53

Chapter 8: The Best Reels, Posts, Wires and Chargers.....56

Chapter 9: Rodeo Stock Grazing Days.....64

Chapter 10: Losing a Lease is Not the End of the World.....74

Chapter 11: Keeping Those Long Grazing Leases.....80

Chapter 12: Reducing Fossil Fuel Consumption.....83

MULTI-SPECIES GRAZING

Chapter 13: Diversify Your Landscape: Multi-species Grazing.....91

Chapter 14: Building Fence for Sheep and Goats.....98

Chapter 15: Those Wonderful St. Croix.....105

Chapter 16: Developing Your Own Parasite-Resistant Flock.....112

Chapter 17: The Economics of Pastured Hair Sheep.....116

Chapter 18: Adding Brush Goats.....120

Chapter 19: Adding Tamworth Grazing Pigs.....126

Chapter 20: Selecting the Livestock Guardian Dog.....135

Chapter 21: Training and Care of the Livestock Guardian Dog.....139

HIGH DENSITY GRAZING

Chapter 22: An Overview of Holistic Planned Grazing.....155
Chapter 23: Holistic High Density Planned Grazing with Small Herds.....159
Chapter 24: Our High Density Grazing Fencing Tech niques.....164
Chapter 25: Our High Density Grazing Water System.....179
Chapter 26: Added Benefits of High Density Planned Graz ing.....183
Chapter 27: Sell Your Mower, Mob Those Weeds!.....197
Chapter 28: Gamagrass Loves Mob Grazing.....204
Chapter 29: Calving with High Density Planned Grazing.....208
Chapter 30: What Will the Neighbors Say?.....213
Chapter 31: The Landowner Mob Move.....216
Chapter 32: Combining All the Herds.....219
Chapter 33: Let the Cows Do the Walking.....224
Chapter 34: Grazing Tips from Ian Mitchell-Innes.....231
Chapter 35: Landscaping versus Animal Performance.....238

GENETICS AND GRASS-FINISHED BEEF

Chapter 36: Grain versus Grass in Your Cow Herd.....243
Chapter 37: Developing or Finding Grass-Genetic Cattle.....246
Chapter 38: Grass-Genetic Bulls.....251
Chapter 39: What to Do with Open Cows.....255
Chapter 40: Fine Tuned Grass-Genetic Machines.....258
Chapter 41: The Marbling Issue.....262
Chapter 42: The Judy Grass-finished Meat System.....264
Chapter 43: The Commodity Side of Raising Yearlings.....270
Chapter 44: My Better Half.....272

Index.....276
Order Pages.....277

Acknowledgments

I would like to thank my parents, Allen and Virginia Judy, for teaching me the value of learning something new everyday. Additional thanks go to my wife, Jan, for all the joy and happiness she has brought to my life. I would also like to thank Steve and Cindy McCawley, Mike Curry, Jack and Dulane Wohlman, Dave and Sue Menius, Monica and John Little, Marshall Colley, Wally Olsen, Ian and Pam Mitchell-Innes, Gearld Fry, Teddy Gentry, Dave Roberts, Kirk Gadzia, Kit Pharo, Troy Bishop, Steve Westhoff, Wayne Russell, Mose Petersheim, Kenneth and Louise Gallup, Steve Landers, Bob Cooper, John Kirchhoff, Neil Dennis, Dave Crider, Bud and Eunice Williams, Allan and Carolyn Nation, Ralph and Jerry Voss, Brent Martin, Frankie Bowen, Sally and Justin Angell, Bill Heffernan, and the Green Hills Grazing Group.

Painting Pictures with Livestock

My name is Greg Judy. My wife, Jan, and I graze grass in central Missouri. We are located 25 miles northwest of Columbia. The area that we live in is rolling hills with clay base and two inches of top soil. It is grazing country, not crop ground.

All though the early 1900s landowners thought they should be able to plow and raise crops here. They all failed after the top soil eroded down into the river and left gullies, which still exist to this day. This is the same condition that most of our farms were in when we first leased them. It wasn't a pretty picture, and most likely, the land you start with won't be either.

It has been a challenge rebuilding the soils and creating pretty landscapes. The exciting part of this soil rebuilding process is that the land is very heal-able with proper livestock grazing management, which I will cover in this book.

I wrote a previous book called *No Risk Ranching* in 2002. It covered finding land to lease, building ponds, developing water systems, Management-intensive Grazing, fencing techniques, calculating the cost of a lease, negotiating land leases, writing land and custom grazing leases, figuring fencing costs, setting goals, controlling costs. This book will not cover those topics but will focus on the new exciting topics that we have learned since I wrote *No Risk Ranching*. Occasionally I will make references to the previous book because of new lessons or mistakes I've made since 2002.

I have had so many people call and visit me since that initial book. They call with many questions or to inquire about touring our farms. I have always tried my best to help answer and educate every one of them.

A lot of them would tell me, "Greg you didn't cover that in your book." Well I have learned a lot more about making money with grass since I wrote the first book. I have also made some more mistakes that I would like to share with you. So that is what this book is going to cover - successes and failures in this wonderful grass grazing world that we have at our fingertips.

Since August of 2002, when my first book was released, we have met more wonderful graziers from all over the world than you can ever imagine. Folks, there are some very competent grass graziers out there who we can all learn from. One thing that is common to every one of them is they have a wonderful attitude about life and are passionate about what they are doing. I don't care what you do, if you love it, you will be very successful at it.

It has been very rewarding to me personally to have a positive effect on some folks who were struggling to make ends meet on their farms. A lot of the people were concentrating all their energy and limited money on the wrong things.

They were doing a lot of the same things we had done wrong before we started leasing land and custom grazing. Some examples were - making hay, owning equipment, selling hay off their farm, not grazing hay fields, no Management-intensive Grazing, being fixated on cattle EPDs, heavy weaning weights, feeding lots of grain, owning land instead of leasing land, not considering custom grazing. The list just goes on and on.

We have learned a lot of new tricks to improve our new leased farms more quickly, and new tips for securing profitable land leases.

The one thing you want to keep in mind while you're reading this book is that this stuff works. We are living proof that it works.

10

There will be snags, but you just work through them and keep your forward momentum going.

You may not agree with everything that you read in this book. That is fine.

Keep an open mind and apply the concepts and methods that you think will work on your operation, discard the rest. You must adopt the attitude that well managed grass is like having money in the bank. If you learn to take care of your grass, it will take care of you also.

There are so many ways to spend money on the farm that it can literally take you over the edge if you are not prudent and regimented with your finances. There is nothing wrong with having an off the farm job, but that should not be supporting your farm life. If that's the boat you're in right now, this book will change that. Your farm must be supporting itself once you quit your job or you may be back in town working sooner than you think.

Whatever you do on the farm must be sustainable, in simple English, "Must be able to perform without any added inputs."

At the present time we are managing seven leased farms and own three. Notice I said own three. I would not recommend this starting out. We used to have two to three custom grazed cattle herds, depending on the time of the year. We now have combined all herds into one large mob. This gives us a longer rest period for the grass to recover, much better animal impact, less labor and a higher quality of life.

What we have been very successful at is "Landscaping With Livestock."

We have developed all ten of our farms with zero tractor inputs. It is so much more economical to let the animals do the work and pay you while they work at landscaping. There are lots of ways to use the cattle as your paint brush and form the perfect painting. We are the artist, the cattle are the brush and paint! Our landowners love our "Livestock Paintings."

We have decided that machinery takes money out of our

pocket. Livestock puts money in our pocket. We prefer the latter method. We have learned how to use livestock to develop and beautify our farms.

We concentrate on letting farms sit idle at key times of the year. This allows us to grow grass and graze year around, and it provides cash flow in the winter months that almost duplicates the summer cash flow. It's nice to get paid every month for your services of taking care of other people's cattle.

Think about this for a minute. We are grazing other people's cattle on leased farms and ours. They are depositing fertility every day they are on these farms and we are getting paid for them while they're doing it! At the same time we're growing more soil and more grass each year. It doesn't get any better than that folks.

The only free input in agriculture is sunlight. We are harvesting every bit of it we can through our livestock.

In 2003 we started purchasing our own livestock for part of our operation to make us more diversified. Custom grazing will continue to be a part of our operation because the cash flow it brings in is awesome. Saving this cash has allowed us to buy debt free our own grass-friendly-genetic cowherd. They perform perfectly on 100% grass. No grain. No hay. No protein licks. I will cover our grass-friendly genetics in detail later in the book.

We started out with 22 cow-calf pairs in 2003, saving all the best heifers each year to put back into the herd. The majority of the steers were grazed up to 800-900 lbs and sold to a grassfed beef company to finish out.

In 2005 we sold our first grass-finished steers in spring, right after grazing them on the spring flush of grass. By purchasing select seedstock and keeping the best heifers, we built up the herd to 280 head of cows, bulls, yearlings, bred heifers, and calves by the end of 2007.

In 2004 we started building our own parasite-resistant hair sheep flock from seven St. Croix ewes. By 2007, we had built the flock to around 300.

I am so excited about our future in sheep. We got extremely lucky and found a fellow who had a few St. Croix ewes and rams that he had never wormed. This was the source of seedstock for our small flock. Our goal is to build this flock up to 500 ewes.

The big problem with sheep in our area is parasites. They are the number one predator in our area where we have 36-42" of rainfall. If you have to worm them, they are not sustainable, and, you will have no profitable future sheep production.

We also have added goats and Tamworth grazing pigs in the last several years, and are very excited about their future in our grazing systems. The goats just love eating leaves off anything with a thorn on it. The pigs love bulldozing through our cow pies with their powerful snouts. Doesn't that sound like fun!

In the spring of 2006 we switched from Management-intensive Grazing to Holistic High Density Planned Grazing on our farms after listening to Ian Mitchell-Innes from South Africa describe his Holistic Planned Grazing system. The results we have seen in two years are just breathtaking.

All of these wonderful results have come without any added inputs, just managing the cattle grazing in a different manner. The future with our farms using Holistic High Density Planned Grazing looks very bright indeed. I do not want to get started talking about it here, or I will not be able to stop. Never have we been more excited about anything in our grazing life.

We also started a website called greenpasturesfarm.net. The website gives information on our meat, books, speaking engagements, and grazing schools held at the farm.

Everything we have bought is debt free, paid for by saving every penny of custom grazing income and paying cash every time a livestock purchase is made. We have eliminated from our operation the process of getting a bank loan, which has been critical in speeding up the process of owning live-stock. No interest of any kind has been paid to the bank. This allows 100% of all income to be available for building our livestock numbers.

It is amazing how fast equity accumulates if there is no interest payment sucking on it each month! If you fill a bucket full of water that has a hole in it, your water runs out through the hole. The hole in the bucket is your interest payment. Put a rubber stopper in the hole and see what happens to the water flow. Before you know it, your bucket is overflowing with your hard earned, saved cash instead of feeding your banker's interest payment.

The only way to control your economical future is to have as much control over it as possible. Money managers, mutual funds, bond managers, stock brokers, the list goes on and on with people who say they can make you money.

We are taking a different route, we are building a diversified, sustainable, profitable ranch. We do not have the worries of a crooked CEO, stocks and bonds crashing, interest rates skyrocketing, and the many other things that can hammer your investments. If we lose equity, we have no one to blame but ourselves, no finger pointing here. We have made the decision that we will control our economic future, not some professional financial company, and it is working very well for us.

One thing you have probably noticed by now is that we are working toward a lot more diversity in our operation. We were 100% custom graziers when we started out. Now as we have grown our equity from custom grazing, other areas of income are being incorporated into our grazing system that complement each other. This is key to growing any operation. Everything that we have added uses the same infrastructure that was in place, other than adding additional wires for our sheep pastures and guard dogs.

No additional overhead was purchased when we added sheep, hogs or goats. Everything that was working with cattle works with the new species as well. I will go into more detail in a later chapter about good grass genetic cattle, but that is the reason we decided to start our own herd of cows. I think we can graze 50% more good grass-genetic, moderate-frame cows on the same amount of grass that currently is being grazed with custom-grazed, large-frame cows. With planned High Density Grazing thrown in the mixture the number we can graze may even be higher than 50%. When you can graze 50% more cattle on the same amount of land, that is like owning 50% more land without buying it.

You wonder why I get all giddy about High Density Grazing. Talk about a competitive advantage folks, we have found it. I'll have much more data on this later in the book.

Have you ever climbed to the top of a hill and pried loose a big rock, then watched it tumble down the hill? Do it. You will notice that the first couple of rolls it makes are pretty slow. The ones following it are faster. Pretty soon you can not stop it even if you wanted to! This is exactly what has happened to us in our grazing company and we absolutely love our life.

Let's go get a few stones rolling!

The Basics

Chapter 1
Education and Attitude: Starting with a Blank Canvas

Let's assume you are starting from scratch with no land. That's actually not a bad place to start.

With no land you don't have a farm payment staring you in the face each month. If I had it all to do over again, I would have never bought a farm and tried to pay for it by grazing owned cattle. I survived, but just barely.

Education and attitude will take you a long way toward your goals in this wonderful world of grazing.

I started out with Management-intensive Grazing (MiG) in 1993 after reading a copy of the *Stockman Grass Farmer* that a really good friend gave me. I have never let my subscription to the *Stockman Grass Farmer* expire since.

This same friend told me about the University Of Missouri grazing school at Linneus. (At the time it was conducted by Jim Gerrish, who coined the term Management-intensive Grazing.) I attended that school as quickly as I could get signed up.

Prior to that I had been continuous grazing because that was all I knew how to do. I was constantly out of grass, fed way too much hay, and had rampant animal health issues. Basically I was doing everything wrong. I was going broke and knew I had to change something or my grazing operation was history. I did not even think of myself as a grazier. I was a cattleman.

If I had not changed my focus and educated myself, I would be out of the grazing business today.

As a result, I started using MiG to set up a grazing

system on my owned farm and all of our leased farms. In 2006 I switched to Holistic Planned Grazing after listening to Ian Mitchell-Innes from South Africa explain his holistic system in Africa.

The first thing you have to do is to get educated in the practice of being a good resource manager. I would highly recommend learning all you can about Holistic Management. The term "holistic" means we are managing for the health of all. With Holistic Management you focus on the importance of giving a helping hand to grasses, bugs, livestock, birds, wildlife, micro-organisms and the whole biological community. When you work with nature you are improving the plants, soils, water table, animal and human life. We are improving the life of all things by concentrating on being good resource managers.

There are tons of materials out there to teach the principles. Visit with successful graziers and learn their management practices. This will save you time and money simply because these guys have made all the mistakes for you. Why go down the road of hard knocks if you don't have to?

When you visit these expert graziers, listen to what they have to say. Don't talk about yourself. You're there to learn from them; do just that. I don't know how many times I've had people come for farm tours and they will start talking about stuff that has nothing to do with becoming proficient graziers. This is rude and it tells me they are not serious about improving their current knowledge. Take written notes as you receive data from them. You will never remember everything they tell you.

I learned a lot over the years after attending grazing seminars and conferences, but I am still not satisfied with my present knowledge. There are always new ideas out there that you can learn to increase your knowledge and bottom line. Keep learning.

Once you're armed with the arsenal of management practices you're ready to go looking for some cheap land to lease. Look your neighborhood over really well. What is going on with the idle land around you?

I had a fellow call me and ask what he should do about a particular tract of land that lay down the road from him. He was visiting with the landowner who started complaining about his pastures growing up in grass. The landowner commented that he was not looking forward to buying a tractor and mowing all summer to keep the grass down.

This is the kind of opportunities that are out there folks.

This fellow called me and asked what he should do. I immediately instructed him to get over there and explain to the landowner how he could fix his problem for free. The 200 acre property had no back fence, but it had three ponds spread over it. The fellow went back to the landowner and secured a free lease on the property in an area where land was selling for $5000 per acre!

There are people out there with enough disposable income to own land and not know how to manage the land.

This is where we come into the picture.

If you become a good resource manager, you will get land, and, you may even have land tracts offered to you. This is because the land you are presently managing speaks volumes about your expertise. You can not hire this kind of work to be done. You must explain to the landowner that you are going to put your heart and soul into improving his farm and you need a commitment from him in the form of a written long term lease.

We have turned down land leases because the land-owner was not willing to give us a minimum seven year lease. I will not touch a farm unless I can get seven years on the contract. I can not over emphasize the importance of having enough time written into the land lease. By the third or fourth year our leased farms are looking so good that the last thing the landowners want to do is kick us off.

Everything collapses without good resource management and the landowner knows it.

Also, with time you have the opportunity to build a good relationship with the landowner. Don't think that you need a large tract of land to start with. Twenty to 40 acres is enough to start with. Learn from this smaller tract. Your mistakes will not cost you as much.

A supportive spouse is essential in the process of leasing land. There is a lot of work initially in getting an idle farm turned into a profitable grazing system. Let her or him know that it is going to take time away from things that you may normally do together to get this grazing system flowing.

In other words, you may have to adjust your schedule a bit to give yourself adequate time to work on the farm. Your spouse must understand this and be willing to sacrifice some time away from you, or better yet, pitch in.

I was very fortunate to find Jan. We went out on our first date in 1997 and I found out real quick that she loved doing things outdoors. When I leased my first farm in 1999, she would come out every weekend and help me build fence, cut brush, work cattle, pound posts, load and sort cattle. You name it and she did it with a smile on her face!

I can remember the first really brushy area that we got into on our first leased farm while building fence. You could not walk or drive up this mountain area of thorny brush. I took a chainsaw and cut out a path along the property line while Jan came behind me piling the brush to one side and painting all the stumps with brush killer to keep them from re-sprouting.

We had a problem. All the posts and wire had to be packed up this mountain by following this narrow path that we had cleared. I did not own an ATV, so we improvised.

The only piece of machinery that I had was a two-wheel dolly, the kind that you use to move a refrigerator. We strapped on the 100 lb roll of hi-tensile wire, posts, and driver to the two-wheel dolly with tie straps.

Jan got behind and pushed while I pulled. We scaled that mountain with our fence building supplies after several stops to catch our breath. We had to do this numerous times to get the back of the farm fenced. You talk about a woman with some grit!

I asked her to marry me in 2001 and she accepted. I am

thankful every day that I breathe that she did. You see, without Jan, none of what has been accomplished would have happened to the degree of success that it has today. Thank you Jan.

Chapter 2
The Toolbox:
The Basics of Livestock Grazing Management

When we first got into leasing land and custom grazing cattle we had no money. Our situation looked pretty bleak.

Looking back, it was a good thing to be broke at the start. It made us look at alternate ways to do things without spending capital we did not have.

If we had had a chunk of money at the start, we would have spent it on the wrong things. One of the first things we would have probably bought would have been a tractor and mower to try and clean up the leased farms. We would have wasted countless hours sitting on a tractor seat mowing, only to see it grow back in a month or two. Instead we had to devise an alternate plan to get rid of the duff and brush without any machinery. Once you're backed into a corner with no way out, you become very inventive.

The first parcel of land we leased was an idle farm that had been abused for many years. Any area flat enough to run a mower and tractor over had been hayed every year until nothing grew back except worthless broomsedge grass. The landowner finally could not get anybody to mow it for hay because the hay crop was worthless.

This farm had not had any livestock on it for 30 years. There was only one section of perimeter fence on the farm that would keep a cow in. Where the old fence previously had been, you could see remnants of old barb wire grown into trees and sticking out of the ground. All the posts had rotted off long ago. The brush in the old fence lines was almost impossible to walk through. All this brush had to be cut before any new hi-tensile electric fence could be erected.

This farm had been offered to two other local cattlemen by the landowner on a lease basis before I contacted him. Both cattlemen turned him down because there was no fence, and a lot of work needed to be done before cattle could be grazed. I knew both cattlemen. Neither one of them was as hungry or desperate as I was.

I was broke and about to lose my own farm. Time was not on my side.

When I first called the absentee landowner, I introduced myself and told him about Management-intensive Grazing. I expressed my interest to him in setting up a MiG system on his idle, unfenced farm. He was shocked that I was interested in leasing his farm because it had no fence.

I agreed to walk his farm and let him know what I thought. I made mental notes of the positive and negative aspects of the farm as I walked over it.

The positive points were:

1. There was a lot of good grass choked back by the duff and brush encroachment.

2. The farm had several undeveloped water sites.

3. The farm was located next to several other idle farms (possible future farm lease candidates).

4. Electric utilities were available for electric fence.

5. There was good road access to the farm.

6. The farm was two miles from my house.

The negative points were:

1. There was poor grass on the best level fields due to excessive hay harvesting.

2. There was no fence to hold livestock.

3. There were no developed water sources.

4. There were no handling facilities.

5. There was excessive brush in the pastures.

After walking the absentee landowner's farm I called him back to give him a breakdown of what I thought about his farm. I explained to him all the negative and positive attributes that I saw while walking his farm. Although the positive attributes far outweighed the negative ones, I made it clear how much work I had looking at me if we were going to develop his farm into a grazing system.

I explained to him a rough sequence of events that would have to take place to develop the grazing system on his farm - spreading lime and seed, fencing it, and getting water ready for the livestock.

I also explained to him that I did not have a lot of money, but that I did have a good working knowledge about how to develop good grazing systems. I told him I would need at the very minimum a ten year lease or I was not interested in leasing the farm.

I went into detail why I needed the long term lease on his farm: It was going to take several years to turn his farm into an average grazing system and additional years to turn it into a great grazing system. I wanted to make sure that I had ample time to reap the rewards of all my hard work in developing a grazing system on his farm. I also explained that it would take three to six months to build the perimeter fence and get cattle on his farm. At the time the cattle were placed on the farm, this would be the land lease starting point. In other words, the landowner would not receive any lease payment until the farm was ready to graze.

Next, I sent him a self-narrated video of my own MiG farm system. On the video I showed paddocks before and after the cattle had grazed them. The video showed close up views of clover and other beneficial grasses. Also included were close ups of the manure distribution. My personal farm basically looked like a beautiful manicured park with all the various legumes and grasses growing in every paddock.

One thing that I would do differently if I ever narrate another video is to make sure I don't have a chew of tobacco in my mouth. Back in those days I still chewed and every time I spit while recording the pasture video, you could hear me spit! It's not the most pleasant sound while you are watching a video. The landowner probably was thinking, "Man, I wonder what hole this hillbilly climbed out of?" Thankfully the landowner never mentioned the extra background vocal sounds of the tobacco spitting that was on the video of my grazing system.

When the absentee landowner watched the video he could actually see first hand what his place could look like some day. It was almost like giving him his own personal farm tour of my grazing operation, except I was in central Missouri and he was in Dallas, Texas.

He loved what he saw on the grazing video and could really visualize his place looking the same with my management.

I asked him if I could write a lease proposal for his farm for him to read over. That would give him the opportunity to voice all his concerns about leasing his farm. I wrote the lease proposal and sent it to him. After some minor negotiations back and forth about the wording of the written lease contract, I signed a ten year farm lease two weeks later on his 160 acres.

I felt like I was walking on water when I got the signed, ten-year-land lease back. I had already gotten a list together of things that had to be done and the order of importance of each item.

The very first thing I did was tackle the perimeter fence on the best pasture section of the farm, which had a small deep pond on it. This would allow me to start grazing immediately. I also found a cheap source of scrap fiberglass posts and used hi-tensile wire for building the perimeter fence. Once the perimeter fence was installed, I fenced off the pond and put a siphon hose with a tank float on it over the dam to feed a 500 gallon water tank.

Next the landowner paid to have an electric meter set so that I could power my electric fence charger. Now that I had water, electric fence and grass, I needed some cows. I put an ad that read, "looking for cows to graze" in the local sale barn. Before I left the sale barn office, a lady who saw me put up the ad asked me if she could come out and look at my grass. She ended up bringing me 30 cows to graze the new leased farm that spring.

Next I started building a working facility out of scrap factory steel siding for catching, sorting, and loading cows. The landowner put down three tons of lime to the acre and I broadcast red clover seed. I had a fellow tell me that I was wasting my time putting clover seed into that old duff. I told him that we had a machine coming that would remove the duff, fertilize and tread the clover seed into the ground with one pass. He looked at me like I was crazy. Of course the machine that I had coming was a group of dry cows that were going to be rotated around the farm.

Dry cows have a very low nutritional requirement in the early stages of their pregnancy, and are the perfect candidate for landscaping on rough overgrown pastures. You do not want to use any young growing class of livestock to develop a farm because their nutritional requirements are not going to be fully met by grazing rank grass.

I immediately started putting in paddocks so we could force the dry cows to eat the brushy material and expose the grass. At the same time we were using a chainsaw and removing all the cedars by cutting and burning them.

In my area cedars are the number one pasture robbers. They will kill out a circle of grass around their whole circumference. Large cedars can kill out a 60-80 foot circle of grass. Even if they don't kill the grass under them, the cattle will not eat the grass growing under them well. The acidity in the ground from all their needle droppings over the years is not conducive to good grass.

Once the permanent paddocks were installed, we used portable poly wire to restrict the cattle into smaller paddocks. It was amazing what the paddocks looked like after several rotations through them. The duff was gone. Left in its place were manure piles and stripped ground cover exposed to the sun. The next rotation through on the same rested paddock was much better. It had new grass plants coming up, the manure had worked into the soil, and baby legumes were coming up.

We had one pond on the whole farm the very first year. We built a series of lanes that hooked to the paddocks to allow access to the water tank. It was not the best, but it got us started in our grazing business. Today that same farm has six ponds and piped pressurized water on five of the paddocks.

The point is that when you first get started, do whatever is necessary to get the cattle grazing. Once the landowner saw how committed we were to developing his farm he built the rest of the ponds right where we needed them.

If at the start we had asked the landowner to build the needed ponds at these locations, he would have told us to go fish! Today, we have 17 paddocks and each has immediate access to a water tank behind a pond. All you need are the basics to get started - water and fence.

After that first grazing season the farm had a whole new look. The pastures were green instead of dead brown duff. The whole place looked alive!

It took some work but the rewards of that early labor are still being reaped seven years later. By controlling the timing and length of grazing periods of the cattle we were able to convert a useless piece of idle dead thatch into a profitable grazing system. In a later chapter I will cover our new improved grazing system that we use to even further produce beautiful landscapes.

I forgot to mention that at the end of that first grazing season, the landowner came over to my house and gave me a cured ham and turkey. He made the comment, "This is a small token of our appreciation for what you have done with our farm this year." Next he reached in his pocket and pulled out a wad of cash. He handed back my land lease payment in cash! I initially refused to take the cash, but he said that if I didn't take the cash he was going to be hurt! So I took the cash and thanked him for the wonderful present.

And, he wasn't done yet; the best present was saved for last. He reached in his shirt pocket and pulled out our written lease. He replied, "After much thought and discussion with my wife we have decided that this written ten year lease is no good."

I instantly went from being on top of the world to feeling like someone hit me in the stomach and knocked the breath out of me.

He further replied, "My wife and I have decided to give you a lifetime lease on our farm. We are so happy with what you have done in one year that we want you to manage our farm for the rest of your life!"

You could have pushed my whole body over with a feather! I could not believe what I had just heard. I wiped the tears that were forming from the corners of my eyes and reached out and gave him a nice firm handshake, thanking him for such an unbelievable gift. I may have even hugged him, I don't remember.

In less than one year my life made a huge transformation. Before I leased his farm in January of 1999, I was struggling to keep the bank from taking my farm. It was common to have $8-10 left in my checking account to hold me over until my next town job paycheck came. I ate a lot of three-for-a-dollar frozen pot pies. Nine months later I had cash flowing from two farms, my own farm and the absentee landowner's farm by grazing other people's cattle.

I had found my unfair advantage, getting paid to graze other people's cattle on other people's land.

Chapter 3
Where's the Grass?
Uncovering the Farm's Hidden Potential

Probably one of the biggest mistakes I've made was completely renovating a pasture that had sod on it.

It's hard to look back and rationalize what I was thinking when I made this decision.

I hired a fellow to chisel plow the existing sod. The old sod was comprised of fescue, weeds, small brush and some moss, and the chisel plow ripped it up into chunks of sod as big as bushel baskets.

Right after he chisel plowed the whole farm we entered a long summer drought. I never ate so much dust and diesel fumes in my life trying to disk out the monster dirt clods. Most were as hard as rocks after being baked in the hot sun day after day. The only way that I got them busted up was by cultipacking the field four times. The cultipacker is basically a very heavy metal roller that has a whole line of big heavy pointed blades that smash the dirt clods into smaller dirt clods.

All the different grass seeds I put down were of various sizes, so this meant I had to go over the entire field broadcasting each grass variety separately with an ATV-mounted grass spinner seeder. Then I had to climb back on the tractor and cultipack the entire field again to ensure good soil-to-seed contact. I was never so sick in my life of breathing tractor fumes and bouncing over the rough broken sod field. At the end of each day I felt like I had wrestled a grizzly bear.

I would have had better results by putting in paddocks, broadcasting clover and letting dry cows do the work. I would have also been getting paid at the same time. It does not take a rocket scientist to figure out which method is more economical. There is not a better cultipacker/soil disturber than a 1000 lb cow's hoof, and the cows would have fertilized it at the same time.

To make money in the grass grazing business you need to have prolific stands of grass and clovers that can take grazing pressure, recover with rest, and start the cycle over again. This takes time and management. Be patient. Keep your billfold in your pocket.

A lot of the farms we have secured leases on were worn out to the point they would not grow good grass. Most were covered with broomsedge, which is a grass plant that comes into fields when all soil nutrients have been exhausted.

Continuously hayed fields are prime examples because the grass was mowed off and removed every year without any nutrients being put back. The soil can only do this for so long until it finally gives up. Broomsedge thrives in this environment, and it makes a solid mat sod that blocks the sun from the soil. Very few livestock will graze it.

By late summer, the field will turn a brownish color. It looks almost like a mature wheat field. One of the first farms we leased was 100% covered with broomsedge from years of being hayed. Finally, the landowner could not get anybody to bale it because the cattle won't eat the broomsedge hay. It is basically cellulose, which does not do much for livestock when the snow is flying.

I remember the late summer day when we finally got the last section of the farm fenced. We turned the cows out in it; they walked from one end of the farm to the other and walked back. They looked at me as if to say, "Where's the grass?

That was a sick feeling to have all the farm fenced and cattle looking at you as if to say, "You're crazy if you think we're going to eat that stuff."

I split the farm into smaller paddocks, then strip grazed it with poly wire to get the dry cows to eat it off. They didn't like it, but they ate it. I gave them a protein lick and they did fine.

By forcing the cows to eat off the broomsedge, they trampled it and opened up the soil bank to the sun. This allowed other forbes and grasses to get started.

The landowner put down three tons of lime per acre, and I bought some 1200 lb big round bales of hay. I started unrolling bales all over the broomsedge. I went through the winter doing this. By the next spring the whole farm was covered with manure and old hay that the cows had tromped into the broomsedge.

I broadcast three lbs of red clover over the farm, and by the following March had clover and grass coming up everywhere. You could see exactly where the hay bales had been unrolled in the broomsedge that spring, and where the best grass was coming up. There were forbs coming up that had been buried for years that had never had a chance to express themselves. The seed from the hay was just a bonus.

We were able to run stockers the following year as grass replaced 70% of the useless broomsedge. Within three years, 95% of the broomsedge was gone.

One of the keys is to remove the duff when you first start, by grazing it.

You simply can not broadcast clover and throw some lime down and hope for the best. The sun has to have access to the soil to bring up the beneficial plants.

Broomsedge loves a monoculture environment. It does poorly when the better grasses start kicking from the added manure and humus that comes from rolled out hay bales.

The method I have described is far superior to completely disking up a field and starting from scratch. It takes too long a time to build up a sod that you can place cattle on without pugging. This lost time is money that you're losing in grazing income.

When you put dry cows on rough un-grazed ground you must restrict them to small areas so that you have a dense mob of them. People tell me all the time, "Greg I don't have access to a large mob of cows."

It makes no difference how many cows you have. Two cows can be a mob if you fence them tightly enough that they trample and eat everything that they have access to that day. Give them just enough each day that it is cleaned up the next morning when you come to move the wire. The cows will be happier if they get to move each day to new rank forage. As long as a cow is moving they are pretty darn content, and stay where you put them.

Also, the smaller the area, the less waste. The cows eat it or trample it. Trampling will build ground litter, which is extremely important. I have a whole section on the importance of ground litter later in the book.

When you move them the next day, if the rank forage is standing, you have given them too big of an area. It must be removed or severely trampled. Remember, you're trying to expose the hidden ground to the sun.

The cow is a very economical machine, learn to use her.

Chapter 4
Our Free Fertility Program: Using Hay to Add Nutrients

Purchased hay is a valuable input on worn out, leased pastures. There are many dollars worth of fertilizer in every big bale.

There is no better method of feeding hay than rolling the bales out over the ground.

I designed a bale un-roller that hooks onto my Toyota or four-wheeler, and un-rolls 1500 lb bales in a snap. One thing that I did not include in my last book was a picture and detailed sketch on how I built it, and a lot of people were disappointed. This book has a photo of it.

This un-roller will unroll a bale over any kind of ground.

People ask me how I can tell which way to hook onto the bale so that it will unroll. With this un-roller, it makes no difference which way you hook on. The hay will come off when you get your speed up.

I personally prefer it when the bale is unrolled backwards to the direction it was made. This gives you a longer windrow of hay, which covers more area. This serves two purposes: it allows more livestock access to it and you get more seed dispersion from the unrolling process. At 20 mph, the bale is throwing off big chunks of hay everywhere, and is fun to watch.

Basically, the way the un-roller works is you hook onto it with a regular ball hitch.

Then you back up to the bale and drive a fifteen inch,

sharpened steel spike into each side of the center of the bale.

These spikes are held in the bale by a chain that is welded to the end of each spike. There is a slot that you set a link of the chain down into to secure the spike and prevent it from coming out in the un-rolling process.

Once the spikes are secured into the center of the bale you can simply unroll it or you can raise the bale up off the ground with the two-ton ratchet come-a-long for transport position. In this position all the weight comes over to the tongue so that all the ratchet is doing is holding it in position until you get to the feeding area.

Once you're ready to unroll, all you have to do is back the ratchet off two or three clicks and this takes the tension off. The ratchet then is removed and you simply take one hand and push the big bale to the ground.

I know you're thinking right now, "Yeah right, push a 1500 lb bale to the ground!"

However, this bale is barely over center and it takes nothing to push it back over center to the unrolling position.

My bale un-roller is so balanced I can roll it around with one hand. It is built like a pendulum. Once you get to the spot where you want to unroll it, simply unhook the ratchet, take your hand and push it over to the ground. The un-roller has a steel skid plate on each side of the bale to ride on the ground once the bale is unrolled. These skids allow the bale un-roller to travel across the ground without digging into your pasture.

In the past, I always fed the bales on the worst areas of the field first, then moved to the better areas later. After attending a Holistic school in Albuquerque in February 2006, I learned to feed hay on the best areas of the field first.

I thought this was crazy when I first heard it. But think about it for a minute. If you feed soil nutrients to the best part of the field and then feed soil nutrients to the worst part of the field, which area will give you the most grass? The best part of the field will win every time hands down.

I still catch myself eyeballing those poor areas and wanting to unroll a bale there. Old habits are hard to break.

I simply do not know what I would do without my cheap bale un-roller. It costs me $250 for labor and steel. I already had a set of spindles for the axles. I bought an electric winch to put on the front to replace the ratchet, but have not installed it yet. The winch will have a plug in that can be hooked to my truck or four-wheeler battery. This will do away with ratchet cranking.

I prefer my Toyota for unrolling hay simply because it has the warm cab in the cold weather. In late winter or early spring when everything turns to mud the four-wheeler is the only way to go. It can get through the mud and not leave ruts like the truck does.

A word of advice for this muddy feeding period, have your big bales in place on the pasture behind poly wire so that you're not trying to transport them through the mud. You will leave some nasty ruts in your pasture, and may even get stuck. By having the bales pre-set, you simply start unrolling them right where they are. Reposition your polywire fence to exclude the cattle from the rest of the bales.

The manufactured bale un-rollers that you see on the back of these big four-wheel-drive flatbed trucks work okay, if you have about $40,000 to throw away for the truck and $5000 for the bale un-roller.

My Toyota is a 1989 model. It is so light it can get over some pretty soft ground that larger trucks would rut up. It gets 22 mpg and the insurance is $200 a year, no payments. If a cow bumps into my truck while I'm unrolling a bale, I don't have a heart attack. A little dent helps build its character. Do not buy yourself into debt. Back to our fertility program.

The cost of nitrogen seems to increase every year. It takes fossil fuel to produce it so I don't think the cost of it is going to come down soon.

I had a young fellow call me the other day. He had put down $9000 worth of fertilizer that spring on his pastures and hay ground. It was late October and he was out of grass. He got only a quarter of a hay crop and still owed money on his fertilizer bill. He commented he would have his fertilizer bill paid off just in time to do it all over again the next March.

This young fellow had a very good job in town but was really getting sick of subsidizing his cattle hobby with his town paycheck. He and his wife had attended our farm seminar and pasture walk. They lived fairly close to us, so they knew we both had similar rainfall totals that growing season. He stated that the very minute they stepped on our farms and saw the grass stockpile that we had accumulated with very little rain, they both looked at each other with astonishment. They could not believe that we did not use any purchased chemical fertilizer and had grown this much grass while grazing our mob of cattle.

I coached him on what he needed to do. Time will tell how well he listened.

Grasses survived and prospered for centuries without commercial nitrogen, so wean yourself off of this profit killer. No hay should be your management goal, but you should have

30 days emergency hay reserve for periods of bad weather. We can buy all the grass hay we want for $15-20 per bale. You can not own the equipment and bale it for that price.

Hay from your pasture could have been direct grazed by a cow that is paying you to eat it. Purchased hay contains nutrients that come from someone else's land and is now going on your land to grow more grass in the future for you. It still amazes me how much technology has been used to take the grass to the cow (mowing, raking, baling, hauling hay) instead of bringing the cow to the grass.

I committed the ultimate sin several summers ago on one of our leased farms. I hired a fellow to come in and bale a bottom of red clover that had gotten away from me. It was about 90% clover. This bottom was a bean field before I leased it. I broadcast clover and grass seed over it. Big Mistake. The clover took over immediately and blocked off the grasses from the sun. This killed the grasses, so all I was left with was basically clover and cockleburs from the bean cropping days. I was afraid to graze it for fear of bloat.

So, I hired a fellow to come and bale it. He charged me $13 per bale to cut, rake and bale it. Later in the year I bought better hay for $10 a bale. To add injury to insult, the clover hay got rained on four times before it was rolled up. I was left with 43 worthless bales that the cattle would not eat and I had to pay $559 to have it baled up. Now, that's really making money!

Once the clover canopy was removed the cockleburs started to explode through the freshly cut clover. I made a promise to myself that I would never again as long as I live make hay on any of our farms. I would have been better off letting the clover get mature, turn brown and let the cattle pick through it on the stem. At least I could have collected a few grazing days from the field and the cattle could have grazed some of the fresh tender cockleburs. More importantly I would not have had all my nutrients tied up in 43 rotten bales of clover hay, but would have been neatly spread across the field on each clover stem and root.

I had to unroll them back on the field to restore the nutrients that I stole off the field when I baled them. Then I was stuck with those rotten moldy bales that I had to move off the field with a friend's bobcat skidloader. The old boy who baled the clover said, "Oh those old cows will eat that stuff when the snow's flying and there is nothing else to eat. They will do fine on it."

Guess what? They didn't eat it.

I have never unrolled 43 bales that made me more sick than those did. First of all unrolling them was like trying to unroll a rock. The white mold inside the bale had the whole bale stuck together like glue. By driving 30 mph with my bale un-roller, I was able to finally get the bales busted apart on the field where they were baled, and I did kill some of those cockleburs by smothering them with the rotten moldy clumps of hay.

The cows laid down on the hay and stepped it into the ground, but never ate any of it. At least I got the nutrients placed back on the field where they came from.

The second method that we use to feed hay is to place the bales out directly where you want them eaten. This method is a lot less labor intensive and works fairly well.

This method works best with 600-800 lb bales. I have a 20' trailer that can haul 12 of these bales at a time. We drive right to the spot and roll them off by hand in groups of three.

We place them on 50' spacing by rolling them. Never leave them sitting on their end. They will absorb every bit of rain that hits them, and this will ruin them for feeding. These 50' spaced bales will be in rows across the field. When it comes time to feed the bales, you can easily move your poly wire back and expose whatever you want to feed. Make the cattle clean them up before exposing any more bales.

The next spring you may have a few small spots where you fed the hay that does not have grass on it. Don't worry about it. The following year that spot will be some of your best grass.

If you use the bigger 1200-1500 lb bales, I have found you get more pugging damage around the bale area because the livestock were there longer eating. If it is extremely cold when you are feeding the larger 1500 lb bales, you will not get any pugging due to the ground being frozen.

Stay away from hay wrapped with polyethylene twine unless you want to take the time to cut it off and remove all the string. With standard twine-covered bales the cows just eat around the twine and it is composted right into the ground.

We like strip grazing off the pastures for awhile, then moving the animals over to a paddock that has hay placed on it. This gives us a break from feeding hay every day and gives the livestock a chance to graze, which I have found they prefer.

It should be your goal to go through the winter and still have some stockpiled forage to graze in late March. In our area stockpiled fescue is king in the winter. The cows absolutely love it, and the fescue sod will support a herd of heavy cows in the wettest weather.

In the winter of 2006 we had 12" of rain in December and 9" in January. Through all of this I rotated the cows on stockpiled forage and kept the cows out of mud. It was so wet that the water just set on top of the ground due to the saturation of the soil.

If I had fed hay during this period the cattle would have been wading in mud up to their bellies. If you can keep the livestock out of mud they perform 100% better. I never doctored a cow or calf the entire winter. I think the lack of mud was a big reason. Once livestock get mud in their hair, they have trouble keeping warm because all their hair strands are covered with matted mud. Can you imagine having to stand in a cold wind with a mud soaked coat to keep you warm? Keeping cattle out of the mud also eliminates most foot problems.

The areas that I strip grazed during the extreme rainy periods had some pugging as deep as 2", but the next spring you could walk over the same rested pasture and it looked like a shiny green legume/grass salad bar. The pugging marks were gone. Nothing but succulent forage was in its place.

I used to think there was nothing better to start the spring grazing season on than winter strip grazed pastures. Everything is fresh, high protein, new growth forage and the livestock just love it.

The transition onto green grass in the early spring is actually better if you have some fall stockpiled grass left over from the winter. The cattle will perform a lot better if each mouthful of grass they rip off has some dry grass mixed in with the new green grass. This dry grass gives them the roughage they need in order to keep their manure from turning to water on a pure stand of green, high protein, spring grass. Otherwise, feed some dry hay while they are grazing the spring grass.

Be on the look out for cheap purchased hay going into the spring. There are tons of it left over by many farmers in our area that they will discount so that they don't have two- year-old hay the next winter. The bales should still be wrapped well with very little settling. If they are about half settled, don't buy them because they will not weather well until the next winter. Offer the farmer a discounted price for the hay and offer to take them all at that price. The farmer has the cash in his pocket and no worry of keeping the hay over until the next year.

I would rather buy this hay than purchase fertility to improve my pastures. You can get paid to feed this hay on your pastures by custom grazing cows. When the weather turns hot and your grasses stop growing in August, you can feed hay onto your pasture and allow the rest of the farm to grow fall stock-piled grass. Remember, you're in the grass selling business. If you can rest your farm and feed hay for a month in the summer, think of all the grazing days you just bought for yourself for the next winter.

Every day that you can cheat old man winter out of feeding hay, you are putting money in your pocket. The winter stockpiling of fall grasses should be planned far in advance. You can not wake up one morning in the fall and say, "I'm going to stockpile grass for this winter." You must have a plan thought out in early summer about which paddocks you will be stockpiling and then graze accordingly.

We try to stockpile our best paddocks because those will give us the most volume of winter feed. We also look at the water systems that service those stockpiled paddocks. We ask ourselves how well the water system will hold up in cold winter conditions.

Every year in October I hear people say, "By gosh, I'm tired of feeding all of this old hay. I'm going to stockpile some grass this fall."

They're too late to stockpile anything for that year. By the next year they have forgotten the pain of feeding hay all winter so they repeat the same mistake again that year.

If you have livestock out there eating it as fast as it grows, it is darn hard to accumulate any winter stockpile in the fall. We have enough farms now that we always have some grass being rested so that we will have something to graze later. It is as good as having money in the bank. It is a wonderful feeling to have a farm of stockpiled grass. It gives you so much flexibility in forage management decisions. It also gives you income every month of the year because you have some grass to sell through grazing.

Chapter 5
Resist Opening Gates During Drought

One of the most devastating things to experience in the livestock business is a prolonged severe drought. There are not to many things out there that are as hard on a person's soul.

Everyday is more of the same, hot, dry weather that further cooks your ground surface. Each day you wake up and pray for moisture to come. The rain clouds seem to go around your farm, and it just seems like it will never rain again.

I remember back in my younger days I made the deathly mistake of opening up gates exposing the whole farm to the livestock when the grass stopped growing during a severe drought. At that point, I was in a death spiral with the future forage production of my farm and did not even realize it.

Going into winter without any pasture to graze was the most miserable time I can ever remember. Every single day I had to feed the cattle purchased hay. I never had the luxury of grazing one single day.

The cattle had very poor body condition by the time spring got there, and the grass still had not started growing due to being grubbed to the ground the previous dry summer.

When I added up what it cost me to take that group of cows through the winter, I would have been ahead money wise to have sold the cows and taken the winter off.

The calves did not pay for wintering the cows. I had a 60% breed back on the cows, and lost two during the winter.

I went through all that misery over the winter and could have sold the cows, thereby preserving my equity, and bought back cows when I had grass. I could have bought two additional cows that spring for what it cost me to winter those cows on hay.

Of course, I was calving in January and February. I was told that was what you had to do so that you would have calves big enough to take advantage of that spring grass when it started growing. What an idiot I was. I listened to the wrong people. Heck, I still have neighbors who calve in the winter and complain about losing calves to the cold every year. You don't see deer giving birth to their fawns in winter. Nature is trying to tell us something if only we will listen. We now calve in May/ June when it is warm and there is plenty of good green grass for the cow.

In the winter of 2007 I spoke at a grazing conference in Northern Missouri. The topic of conversation among a lot of the attendees was how many calves they had lost due to the extreme cold weather. That particular day, it was 10 below zero F outside with the wind chill factored in. Calves are born with their summer coat on. How would you like to be dropped onto bitter frozen ground with your summer clothing on?

One lady had five calves on her back porch under heat lamps, bottle feeding them to try and keep them alive. This is what happens when you go against Mother Nature. Sorry, I got off of the subject of opening gates in a drought, but I had to get that winter calving thing off my chest.

Never at any time should you open all the gates to all pastures and let the livestock eat what they can find. You will be much better off culling heavily or de-stocking the farm rather than letting them eat every pasture you have into the ground. I would rather feed hay for one month on just one paddock. Granted, you will get overgrazing on this one pad-dock, but this is far better than overgrazing the whole farm. Cull your herd heavily and early to reduce the hay budget.

Once you reach the point of opening gates during a drought, you have officially given up any near term future of growing any forage on that farm for the rest of the year, and

quite possibly the next year as well. It will always have devastating effects on your pastures. It may take several years to get
your pastures back to where they were before the drought hit.

What happens is that the livestock remove what little
grass forage is left, stripping the last amount of energy that the
plant had stored for re-growth when growing conditions returned. The plant is in a survival mode. It has to stop sending
out plant tissue or it will exhaust it's root reserves and die.
Once the root reserves are exhausted, your pasture is dead for
that growing season. You may even kill large stands of your
grass.

The ground will become brittle and crack open exposing it even more to the drying elements. Once you remove all
surface litter, you have no mulch to catch any rain and absorb it
in the ground. The surface litter also helps keep the soil surface
protected from the devastating effects of the sun. If you leave
your pastures with some residual showing during a drought
they will jump back when the rains return. We are dealing with
a solar collector after all. If you take off the solar collector
panel how much energy can it produce?

I remember the sick feeling in my stomach as winter
approached after I had opened up all my gates and let the cattle
grub down all the remaining grass. If it is late October and
you're already feeding hay, you are in for a long, miserable
winter of costly mandatory hay feeding.

There is a huge difference in feeding hay and having to
feed hay. You never get a break when you have to feed hay
every day of the winter. Occasional hay feeding interspersed
with grazing stockpile allows you to breeze through the winter
period pretty much pain free.

When it comes to managing during a drought, this is
where custom grazing comes in so nicely. You can have the
livestock removed, because you don't own them. Always have
this clause in your grazing contract - that you will give the
cattle owner 30 days notice to remove his livestock in case of a
drought.

Learn to adopt the philosophy that "Drought is normal. Manage for it." Custom grazing also makes it painless to restock after the drought is over. Just call up your cattle owner and tell him you need some cows. He will jump at the chance to bring you some cows to graze because he is sick of feeding them hay.

Another drought tool we use is to stretch out the recovery period of our grass from when it was grazed last. You must give the grass a longer recovery period to grow back simply because there is no moisture for things to grow. By slowing up your rotation through the paddocks, this will give them more time to re-grow.

Combine your multiple herds into one herd. This gives the other areas of your farm time to grow back.

Another drought tool we use is to stockpile excess grass in spring for summer drought periods. You always have grass that gets ahead of you in the spring and puts on a seed head. Let the paddock set until the summer drought hits, then strip-graze it while you are allowing the rest of your farm to rest during the drought.

I can not count the number of times that I have seen people mowing off their spring pastures to remove the seed heads. One month later they are feeding expensive hay. Where is the logic in that?

I'm convinced that a lot of people get all caught up in the travesty of the almighty seed head. I would much rather have my cattle eating a plant with a seed head on it in a drought than having to feed them an expensive bale of hay. Back in the old days when cattle took care of themselves, do you think they avoided a seed head in a drought? I kind of doubt it. All of these grazing tips will extend the time needed for your plants to recover during a drought.

I hope I've convinced you not to open the gates. It cost me dearly. The one thing that still sticks in my mind is the daily drudgery of trying to keep the cows fed and hoping for things to improve. It's very hard on you mentally to go through this period. Spring seems like it will never arrive. Then when spring arrives, your pastures act like they are still asleep because of the beating you gave them the previous fall. This really leaves a sick feeling in your belly.

The people who did not graze their pastures into the dirt during the drought are grazing the new spring grass. Be one of these people.

Chapter 6
Controlled? Burning

There may come a time in the future when you will feel like you want to burn a field off. I am going to do my best to discourage you from doing that.

Burning increases plant spacing, and a host of other detrimental things result as well. You can kill all live grass roots up to three feet deep by burning off a heavy thatch that burns hot.

Now if you're in the grass growing business and you kill off all your grass roots three feet deep, how much sense does that make?

By burning off the thatch, you are exposing the ground surface to erosion due to lack of ground cover to hold the soil. Once an erosion ditch starts, it is tough to seal it over. Burning also kills all microbial activity in the soil surface. If you're wanting to grow grass, that's not exactly the way to go about it.

All living bugs are torched as well. The potential humus that was covering the ground is deposited in the air. What a waste. Instead of feeding your soil microbes the ground litter, you just torched it into the air.

Some of you who read my first book are probably scratching your head because I am contradicting myself from what I said earlier. Well I have learned the negatives of burning over the last several years and that is why I have changed my tune. Yes, burning does make a good seed bed for legumes, but so does High Density Grazing, which I will cover in detail later. With the grazing method all ground cover is held in place and deposited for the soil microbes.

Every time I burned the gammagrass in the early years, I held my breath that the fire did not get away from me. I will never forget the first year I burned it off.

We lit the fire on the down wind side, forcing the fire to burn into the wind. Next we started bringing a line of fire up the side of the field to light the head fire. It was about dark when we lit the head fire. The head fire absolutely exploded and sent flames 15 to 20 feet in the air as it stormed across the dry thatch field. The whole bottom looked like a football stadium with night lighting, scary sight to say the least.

I had solicited the help of my new neighbors who had been doing some spring fescue burning on their farm. I don't think they had a clue what they were getting into when they accepted! The husband commented to me after I lit the head fire, "Greg you don't mess around when you burn!"

I tried to act like the old veteran and replied, " Kind of pretty ain't it?" I had to act like I knew what I was doing or they would have dropped their rakes and run!

A word of caution on burning gammagrass if you decide to. Have a pressurized water source on an ATV and plenty of help. Once you light the match, you are responsible for the fire and whatever it burns you pay for out of your pocket. This includes houses, barns, heaven forbid, even people. It is a very helpless feeling watching a fire of this magnitude knowing that you could not stop it even if you wanted to. It produces a roar like a train, and it really is a monster.

That gamma fire that night was so hot that it burned off fiberglass fence posts. It also took the coating off the high tensile electric fence wire, which rusted later.

Another burn experience we had was on a leased farm that we had just secured. We had been cutting thorn trees in the early spring and burning them in piles. We had a bed of red coals that was two feet tall. It was late winter and not much green showing yet. The soil was fairly moist from a recent rain so I felt safe burning. We had a burned area around the coals about 20 feet out on all sides.

Everything was fine and under control. We decided to take the dogs down to the creek for a drink and a swim to cool them off. We were gone about 15 minutes and when we returned the fire had crept through the burned area somehow and was trickling through the old blackberry vines.

I immediately knew we were in trouble and jumped on the ATV and drove like mad back to the truck to get two 5-gallon buckets of water and big towels. A big wet towel is an awesome fire fighting tool. I far prefer a big wet towel over a rake or shovel any day when fighting fire. When I got back the fire was headed down two ridges. I told my wife to take the west ridge and I would tackle the east ridge.

We were fighting a head fire that was running through briars. It was hard to get a good swing at the flames. Jan's fire kind of sputtered out once it hit a grazed off pasture, but mine was down in old accumulated duff and had plenty of fuel.

It jumped a 30 foot creek and was headed for a 20 acre gammagrass field that had four new creosote-treated utility poles running through the middle of it. All I could envision was those poles going up in flames like roman candles! At $1200 a piece to replace the utility poles, I was sweating bullets. My bucket of water was gone. I was beating flames with a dry towel now, with no time to run back for more water. I had to stay with the line of fire. I felt like I was gaining on it. The fire was starting to lick the edge of the gammagrass field when I beat the last flame out with my half burned powder dry rag.

I have never been more physically exhausted in my entire life than at that moment. I collapsed and gathered my strength to walk back up the hill to the ATV. When you have a runaway fire, there is no time to catch your breath. Fire waits on no one! My wife and I both looked like we had been beaten with a thorn tree and were completely black from head to toe. Our clothes were torn to shreds, not a pleasant experience at all.

About that time we heard the fire truck sirens coming down the road.

Panic struck. I immediately thought that the fire had erupted again and somebody had called the fire trucks. Our neighbor had decided that he would do some burning as well that day. He was not as lucky as we were. He burned off his farm and part of his neighbor's. If the fire trucks had not arrived, he would have burned off a whole section of land, fences, buildings and all.

I have several more burn experiences that were kind of hair pulling as well. Bottom line about burning pastures is that it is dangerous. If you do decide to burn, please be extra careful. One bad fire experience may put an end to your grazing business.

Burning is not worth the risk that is associated with it and for the results you get from it. Why not let the High Density Grazing method replace it? Your pastures and soils will thank you with strong swards of grass along with fat cattle. You will also sleep well at night not worrying about wild fires burning up the neighborhood.

Chapter 7
Why Switch from Stockers to Dry Cows

Over the course of the first five years of custom grazing we grazed primarily stockers that ranged in weight from 400 to 600 lbs. We got paid by the amount of weight that they gained during the time they were under our care.

I enjoyed watching young stockers put on weight as the summer progressed, but I felt like we were not improving our forages adequately. A lot of time was spent initially with each group to make sure we got them doctored when they got sick.

After about a month of being on our farm, we hardly had anything get sick, and could concentrate on grazing them. It was hard to get any kind of animal impact with stockers without sacrificing gains on them. If you force a stocker to clean up a pasture, your gains plummet.

A dry cow is a cow that has had her calf weaned off and is bred back. Dry cows are our animals of choice now.

They are a very versatile tool to rebuild your soils and build a very profitable custom grazing operation.

When we get our dry cows on the farm they are usually pretty thin. The calf has been weaned off, which takes a tremendous load off the cow's nutritional requirements. Dry cows are a lot less work to take care of than stockers. They hardly ever get sick. You may have an occasional pinkeye or foot rot problem to deal with, but you don't have to be concerned with shipping fever, pneumonia, scours or other diseases.

Dry cows are a lot less finicky about what they eat as well. If you stock them dense enough, they will eat about anything that doesn't eat them first! It still amazes me the body weight condition they can put on when grazing some pretty rank forages. A stocker calf would dwindle away to nothing on the same forage that a dry cow gets fat on.

Another bonus for the dry cow is the actual weight of the cow creates wonderful animal impact. A small stocker does not condition the ground like a 1200 lb cow does. The cow's hoof with all that weight on top of it will break up litter, open capped soils, bust fallen branches, destroy young brushy trees, fill in erosion ditches, and create many other benefits. It takes a heavy ruminant animal to wake the soil up. No lightweights need apply for this job!

I've seen a dry cow mob go into a brushy area, and 12 hours later it looked like you took a fire torch and burned every leaf off as high as you could reach. One of our farms has several old bulldozer piles on it. This is the first place they go if a pile is included with their grass strip. The forbs and bushes that grow out of these piles are absolute candy to the cows.

I saw an old cow actually walk a big old log like a goat to get her tongue around a precious forb that was just out of reach. She did this immediately after she was turned into the paddock. Grass was everywhere, but she wanted that succulent forb! I wish I'd had my camera with me to take that picture. It was an unusual sight to see a huge cow acting like a goat.

Most cows are smarter than stockers. They figure out what you want them to do very quickly. Our cow mobs will see us coming down the road from a distance and they are immediately gathering up by the fence to be moved. They know our truck, probably by the rattle noises!

Cows as a whole are easier to keep in than stockers. We never had a cow out of their mob the whole year. I think with mob grazing they get so accustomed to being in a tight group that they are very uncomfortable emotionally if they leave the group.

Most of our farms were pretty worn out when we leased them. Some of the farms still have a long way to go to catch up with our better farms. This is another reason we switched primarily to cows. Our worn out farms are better suited to cows than stockers. These worn out farms are getting better very quickly now that we have switched to High Density Planned Grazing.

You have to realize what your strengths are and concentrate on those areas. Our strength is lots of idle worn out land that is perfectly suited for dry cows, not stockers. The really exciting part of this whole system is being able to physically see the changes in our pastures in one year.

These custom grazed dry cow mobs are making us more money every year because of the increased grass growth.

Before, with stockers, we were utilizing about 50-60% of the grass with each paddock move. The remaining forage was left standing, not even trampled. Stockers avoided weeds like the plague. Now with twice daily moves with dry cows, we are utilizing 70 to 90% depending on the growing conditions of the grass. Every square inch of the temporary paddock has had a cattle hoof pound it, which gives us the ground disturbance we so desperately need to improve our forages.

Chapter 8
The Best Reels, Posts, Wires and Chargers

We have developed a fencing system that works great for us in our High Density Grazing system. It is simple once you understand it.

This system would also work great with any other rotational grazing system. We are still coming up with new ways of doing the daily or multiple daily moves.

I am going to try and explain to you the best and most efficient methods that we have found that work on our operation. You may scratch your head after reading about our methods and come up with a better one for your operation.

It typically takes us about five minutes to take down a fence and five minutes to put up the next one.

There are several items I want to address before discussing our High Density Grazing techniques in later chapters.

The very first item is time.

If a temporary paddock takes a lot of time to put up and take down, you will get tired of putting up paddocks, and after a period of time, quit. So we are going to concentrate on making it easy, fast and painless.

The next item that we need to discuss is polywire reels.

There are many different reels that you can choose from. I have made this mistake, choosing this reel and that reel from different manufacturers. Every one of those reels are hanging on a rusty hook in my shed.

The only reel made in the world that will perform on a rigorous daily basis that we have found is the O'Briens 3:1

geared ratio reel. Why in the world the other reel manufactures will not wake up is beyond me. The O'Briens is made in New Zealand and has been used many years over there by thousands of graziers. I guess we should call it the John Deere of reels! If I had to use the other old reels hanging in our shed, I would not be High Density Grazing. That is a pretty strong statement, but that is how bad they are in comparison to an O'Briens reel.

O'Briens makes several different types of reels. Make sure you get the geared 3:1 ratio. I tried a couple of their 2:1 ratio reels because they were narrower and seemed a tad bit lighter. Don't use them, they are a pain to reel up wire. The gears on them are very tight and will work you to death rolling up polywire.

They sell a reel called Mega Reel. It is a good reel for rolling up poly tape. It is much wider than the standard reel. They also sell 1:1 ratio reels for whatever reason. They could save on their overhead and scrap those suckers. Do you know how long it takes to roll up 1600' of polywire with a 1:1 ratio reel? I can sit down eat a sandwich and drink a soda in the same amount of time. The 3:1 ratio reels spin three times for every turn you make with the handle.

All O'Briens reels carry a five year, all weather guarantee as well. The O'Briens reel also has a plastic guide that helps prevent the polywire from jumping off the spool and getting around the spindle while you are reeling it in. You notice I said, "it helps." The wire can still jump off if you are not paying attention while you are rolling it in. The plastic guide is a huge improvement over having nothing to guide the wire on the reel.

Now that I have told you all the things that I like about the O'Briens reels, let me tell you a couple things that I do not like. Maybe if an O'Briens person reads this, they will correct these problems.

The first thing that needs to be changed is the plastic guide material. These guides will break sometimes if the reel is accidentally dropped right on the guide itself. I have had them break while rolling out the wire behind the ATV if the wire gets jammed on the reel.

The plastic clips that hold the plastic guide onto the reel are not very durable. Once the clip snaps, your guide will not stay in place. The guide needs to be made with a tougher plastic or beefed up some. These guides are $5 each to replace.

They do make a steel guide that fits the standard 3:1 ratio reel that is more expensive. The steel guide will add more weight, but is durable.

The lever that holds up the locking mechanism on the reel is not very good. You can be driving along letting out wire and all of a sudden the plastic lever lets the locking bar fall down. Before you can stop, the polywire gets snapped in two. Now when I buy a new reel, the first thing I do is take off the whole locking mechanism. You don't need it anyway with our method of putting up fencing systems. If it would lower the cost of the reel, I would suggest O'Briens leave off the locking mechanism.

All right, I have done enough reel re-engineering. Let's talk about the temporary step-in posts.

Here comes that O'Briens name again. O'Briens makes a 42" tread-in that comes in multiple colors that is superior to any other post that I have tried. They have a high tensile steel spike on the bottom that keeps the rust off of the spike. The diameter of the spike is small enough that you can push it into some very hard ground. I have pushed them through two inches of ice-covered ground using the correct technique.

We have a pile of old tread-ins from various manufactures sitting in a pile under those old reels in our shed. Every one of the old posts has broken hooks, rusted spikes, or are bent and broken. Don't waste your money on any other post. Buy what works.

We have 400 O'Briens posts, and I have broken one. The county road mowing crew hit it with their mower! It was my fault. We accidently left it along the road when we built a temporary cattle lane.

We use the white posts because the cattle and wildlife

can see them. In the winter we have not had any problems using white posts on snow, although the black, yellow or blue would show up better. The nice thing about these polyethylene posts is that you can grab an arm full of them and put up a quarter mile of fence with no problem. They are super light, unlike the steel shafted pigtail step-in posts.

If you grab up an armload of pigtail posts, you will have quite a time keeping them under your arm while you are laying out the fence. They slid around, are heavy, and get tangled up with each other. They are also conductive if you accidentally hit the wire against the steel shaft.

We do use one style of steel shafted pigtails for making corners and end posts for our temporary polywire paddocks. Guess what kind it is? Yes, O'Briens just came out with a red (wish it was white) pigtail post that is 36" long. When you press it into the ground it is only 30" tall. This is a full 6" shorter than any other standard pigtail post on the market. This 6" in height difference makes it a very sturdy corner post to pull against or make a corner for a lane. The shorter 36" post gives it higher cantilever strength than the 42" post. People can not believe it until they see it with their own eyes, but it works wonderfully.

This one post makes our whole system much easier. Think about it. If you had to drag a steel post driver and a separate solid corner post for every paddock that you put in, that turns into work real quick. You simply do not have enough arms to pack it all.

I had a fellow come out to see our system, because he could not get it in his head how we were effortlessly putting up all these paddocks. After he watched me put in an O'Briens red pigtail for a corner post he replied, "That's cheating!"

Well, it may be cheating, but it works great.

You are probably thinking by now, this guy works for O'Briens and probably is getting a bonus for pushing O'Briens products. No, I just report what works and doesn't work on our farms. There are tons of products out there that claim to do the job, but we have found what works.

One other post that I want to mention is the Powerflex post. It is a wood plastic composite post made out of 30% wood fibers, 70% polypropylene. We put in a perimeter fence with them and have been well satisfied. They have a UV inhibitor in them and carry a 15 year manufacturer's warranty. You can drive them with a standard post driver. Drilling them is a snap after you have driven them in the ground.

The posts are almost impossible to pull out of the ground once they are driven. They appear to swell up against the soil when driven. You do not need to worry about them being pulled out by something running through your fence. They are the most flexible permanent post on the market. You can bend a post completely to the ground and it will pop right back up to its position.

The posts come in many different colors, lengths, and diameters to suit your fencing needs. Another advantage of these posts is that there is no painting or fiberglass splinters to worry about. Dave Krider and Gary Duncan own Powerflex Fence Company. I have not found another fencing company that carries such a wide assortment of good supplies. They buy everything in bulk and pass the savings on to the customer.

They are both seasoned graziers and use these items to make sure they perform like they should before they place them in their catalog. We were pricing hi-tensile wire several years ago all over the Midwest. For a 4000' roll of 12 gauge wire everybody was asking $100 to $130 per roll! Steel prices had increased substantially and were still going up. I gave Dave a call and they priced a roll to me for $69. We bought 30 rolls, which was all we could get on our truck. Dave commented that they have been accused of cherry picking the best products from multiple manufactures. I say good for them. Selling the best from each company just makes good business sense to me. Check them out on their website folks - just Google the name. They know what they are doing and are great people to work with.

No, I don't work for Powerflex either. I have no problem bragging about someone who is the best in their field. Maybe Dave will buy me a steak when he reads this!

The polywire we use is always white. I used some other colors and had problems with deer and cattle going through it. They simply could not see it until it was too late.

I remember buying a roll of orange polywire once. It was so pretty and bright, no way a critter would go through that stuff. I found it the next day in some of the nastiest balls of knots you've ever seen spread all over the farm.

Our standard polywire has six steel filament wires wound in with the poly string.

A salesman talked my wife into buying a roll of Turbo polywire one day. This wire has nine steel filaments in it and is a lot thicker than the standard six filament polywire. It was also 35% more expensive, and the length of the roll was much shorter.

I don't like it because it fills up a reel so fast with a shorter length of polywire. Granted the nine filaments are a lot more conductive than the six filaments, but nine filaments are not required to keep in livestock with constant moves to fresh grass. Some people use polytape to run their temporary paddocks. Same problem - you can not get very much tape on a reel compared to six strand polywire. Plus, polytape is much more expensive. It is, however, very visible.

When it comes to repairing broken sections of polywire there are several ways to do it.

The method described by the polywire companies is very time consuming. What they recommend is melting the poly string back several inches, then twisting the steel filaments together, followed by a knot.

We keep it simple. We just tie two plain knots side by side in every break that we have. Two knots will give you a better conductivity than one knot. We have one reel that must have 50 knots in it. This is a result from the early days of not having enough reels. I would run out a paddock, tie a handle on it, and cut the wire. Then I would go repeat the same process with the same reel on the next paddock.

Even with numerous knots along the full length of it, the wire will still knock your socks off if you touch it. I held a digital voltage meter on this reel and measured the voltage as I walked down the fence. The drop in voltage was 1000 volts. It was 7000 volts where it hooked onto the source, 6000 at the far end of the field, not bad for a reel that had over 50 knots in it.

There is also a new product out that is braided twine. It's awesome wire, very stout, but very pricey. I would have to sell a couple of our cows to fill all of our reels with this braided twine. I guess it all depends on how deep your pockets are!

A good charger to power your fence is the most important part of the electric fence system.

I never buy cheap chargers anymore to try and save money. You notice I said, "anymore." Well, starting out I did buy several cheap chargers to power some of our farms. It cost me dearly in money, aggravation and time wasted.

The first cheap charger that I bought worked about two weeks. We had a thunderstorm come through one evening after the fencer had been installed. The fencer was not popping the next morning. After a close inspection, the fuse that was mounted inside the box was blown. I thought, no problem, replace the fuse, and I will be back in business. I replaced the fuse, but the fencer still would not work. The charger was still under warranty. The circuit board was fried.

This fencer had a lightning arrester and lightning choke installed to protect the fencer, but it didn't help. I got the repaired charger back, and every time a rain cloud passed over the farm the fuse would blow. I'm being a little sarcastic, but seriously this charger had to be constantly checked for blown fuses. I had to keep a box of fuses with me at all times to keep the charger working.

I am not going to tell you what brand of charger it was because I don't want the company mad at me. I will tell you the name of the charger that I replaced the cheap one with. Stafix chargers are the best charger made in the world, as far as I am concerned. They are expensive, but worth every penny that you pay for them. When you pick up a Stafix charger and compare it to any other brand, it is like comparing a bicycle to a Lexus. I have not found anything that I don't like about a Stafix charger.

We bought the M36 model, which is the largest size they carry. It has 54 stored Joules and 37 Joules output to the fence. It is capable of powering 250 miles of fence. We bought the remote control option that also is a fault finder. We can touch the fence with this remote control and turn the power on or off.

I will never forget the first time I reached out with the remote control and turned the power off. The remote digital screen said it was off, but there was no way I was going to reach out and touch it. The fencer was putting out 9000 volts over four farms when I shut it off! I took my other old fence tester out and it read 0 volts. I was still a little bit scared to touch it. What if the fence came back on by itself for some reason?

I do not understand how the remote works, but it sure is handy to be able to kill the power by just touching the fence and pushing the off button. The fault finder has arrows on the digital screen that point which direction the short is. This is a major time saver in locating the problem short on your fence. The charger has never failed us yet. The fencer has been running three years now without being unplugged. Nothing bothers this charger.

We have brush and weeds all over the fence. Nothing bothers it except grounded steel. No charger can handle grounded steel. The charger has ten 6' ground rods on it for a grounding system. These ground rods are driven around a pond for constant moisture conditions. No matter how dry it gets, our grounding system never fails. If there was a better charger out there than a Stafix, I would be using it.

I should have three steak dinners coming, one from O'Briens, one from Powerflex and one from Stafix!

Chapter 9
Rodeo Stock Grazing Days

We're going to take a little break here so that I can tell you about my rodeo stock grazing days.

Years ago when I was single and first started custom grazing, I was hungry for any kind of critter to custom graze. I signed a custom grazing lease with a rodeo outfit for grazing their bucking horses. The rodeo owner paid me 20 dollars a month for grazing each horse. I really felt like I had hit the big time. That check every month felt like a million dollars simply because I was broke. I had never had anybody pay me that much money to graze anything.

The horses were only on my property five days a week. The rest of the time they were at the rodeo. That amounted to 20 days grazing every month or $1.00 a day. Man I thought I was getting rich!

They had 30 head of bucking horses. Sometimes their numbers got as high as 45 head of horses. On Monday mornings after they returned the horses from a weekend of bucking, it was hard to tell what kind of horses I would find in the pasture. They would actually jump the bucking horses off the five-foot-deck opening on the semi-trailer directly into the gateway of my pasture.

The rodeo owner loved to trade horses, which meant that we always had new horses coming in. Heck, they even turned out two stallions with 30 mares one time.

That was quite a battle to witness. The two stallions would face off, rear up, grab each other by the neck with their teeth and fall to the ground screaming.

I was scared the two stallions were going to kill each other. I called the rodeo owner to tell him about the battles they were having. His reply was, "Heck, son they will work it out in a couple of days!" Those rodeo folks are just a little different breed. They're tough people, who have different ways of doing things than most of us.

Those two stallions each soon had a harem of mares, and each stallion constantly kept his mares grouped together. If a mare got too far out of their group, the other stallion would try to steal her for his harem.

They were big horses. Most of them had some draft horse breeding in them. The larger horses held up better in the grueling rodeo circuit.

The deal was that I grazed them through the week and on Friday night I would help catch them to load for the rodeo that weekend.

Every Friday night I had my own rodeo helping the rodeo boy catch and load the bucking horses.

I would saddle up my fox trotter mare and the rodeo boy would try to ride anything in the pasture that he could catch. He never brought a saddle or a bridle, just a halter and rope. This kid was tougher than nails and built like a bulldog. When you looked at him all you saw was "tough." All of his front teeth were gone, and he had scars all over his face. He always had cuts, bruises, bandages, or an arm in a sling when he arrived at my farm. Some of the injuries were from the rodeo, but most were from fighting after the rodeo.

Those guys liked to party and mix it up after the rodeo was over. They would all meet behind the bucking chutes and start drinking beer to celebrate. A fight was guaranteed to break out. Those guys are just plain tough cowboys.

The rodeo owner sent a different fellow up one weekend with the semi-trailer to help me catch and load the horses. This dude was really banged up. He had a broken collar bone, broken nose, broken jaw, and a swollen face. He had a bandage around the top of his head, a cast on his left arm, and walked with a real bad gimp. He also informed me that he had lost his memory. He looked like he had been run over by a huge truck.

I cautiously asked what had happened to him.

He commented that the last thing he remembered was diving off his horse and grabbing hold of a 700 lb horned steer. The first thing he could remember after that was waking up in the hospital on a Sunday morning. They kept him in intensive care until the swelling in his brain went down.

This guy was a professional bulldogger. Bulldoggers run their horse up against a big steer, jump off and grab the horns, and give a hard twist to the steer's head to throw it on the ground. The cowboy who does it the quickest wins the event. Sounds like fun to me!

Seriously, this cowboy was built like a barn. He had to weigh 280 pounds, not an ounce of fat on him. He released himself from the hospital and drove that semi rig 200 miles to my farm. All the guy could talk about was the next big rodeo in a couple weeks, and how he had to get himself healed up by then.

I think those guys are just plain addicted to rodeo. They've told me that there is not a bigger head rush in the world than coming out of the chute on a 2000 lb mad bucking bull that is trying to kill you. I will take their word for it!

Back to catching horses.

Rodeo boy and I would ride out to the pasture where the main horse herd was and start moving them toward the corral. The boss mare would take off running as hard as she could down in the brush, taking the whole herd with her.

Rodeo boy and I would spread out on our mounts and try to drive them out of the brush. Those bucking horses were escape artists when it came time to rounding them up. The horses knew that Friday night meant going to work. They just were not real happy to comply.

After about an hour of chasing the bucking horses back and forth we would finally catch them. Next we had to load them onto the semi-trailer. We loaded the heaviest horses into the front compartments, the lighter ones into the rear of the trailer.

I've saved my best rodeo story for last.

It was Friday load night. I was sitting on my horse waiting for rodeo boy to arrive. Somehow I managed to catch all the horses by myself and was ready to load them. Well, the rodeo owner sent another driver helper in place of rodeo boy. This guy was from Australia. He spoke with a definite Aussie accent and was one tough looking dude. He had scars and bruises all over his face. I saw one tooth in the back of his mouth when he talked. I noticed when he pulled the semi-trailer up to the loading alley that the bottom (belly) part of the semi was packed full of huge bucking Brahman bulls.

The Aussie also had one great big monster bull on the top compartment by himself. I kind of wondered right then why that bull was by himself.

The Aussie replied, "Give me a hand mate. We got to take off the top bull before we load the horses."

I immediately responded, "You ain't taking that bull off that trailer. My little flimsy corral will never hold that monster."

The Aussie responded, "No worry mate. He is a jolly good fellow."

I reluctantly agreed to let him unload the bull in my weak corral. Before unloading the bulls we sorted the horses so that we had the large horses ready to load first.

I will never forget the sight of that huge bull squeezing through my puny cattle panel alley to reach the corral pen, which was also made of cattle panels. The bull had to weigh over 2500 lbs. He was a very tall, wide, deep animal with a very large hump over his front shoulders. He had two large curved horns that looked like he could gut you with one swipe! He just had a mean look to him with those big old ears hanging down by his eyes.

Once the bull reached the front pen the big horses

started chasing him and biting him.

I immediately thought, "This is not good!"

The bull made about two rounds around the corral with the horses chasing him, and the cussing Aussie trying to beat the horses off the bull with a club. About the third round the bull made a new gate in my corral.

I was horrified. This bull looked like an elephant lumbering down into the woods. It was almost dark by then, and after the Aussie got done cussing he replied, "Let's go get him."

I'm thinking to myself, "You go get him!"

Light was fading fast, and I did not want that bull loose on my farm. So the Aussie and I devised a plan to run him into the corral. We quickly loaded the horses so that the corral would be empty for the bull if we were fortunate enough to get him in it.

The Aussie was to circle around the woods and run the bull out towards the corral. I was to act as a blocker. If the bull came south, I was to head him off. We were both on foot, because the brush was so thick that you could barely walk through it.

All of a sudden I heard a blood curdling scream from the Aussie. The scream was followed by a whole bunch of Aussie language that may have been some form of cussing. This Aussie was flat making a racket down in those bushes.

Well, when I had taken my position for blocking the bull's escape path I had brought along a weapon. I had a solid 2" fiberglass fence post about five feet long for persuading the bull not to come my way!

You are probably thinking by now, boy that Greg is pretty darn brave, or stupid. In hindsight I will vote for the latter. No, I was scared to death, but I remembered what that Aussie had told me. He said the bull was a "Jolly good fellow." So I'm standing there in the dark with this club silently talking to myself saying, "He is a jolly good fellow. He is a jolly good fellow." The other silent voice inside of me was saying, "Why was that bull on the top compartment by himself if he is a jolly good fellow?"

About that time I heard a loud bellow from the bull. The Aussie was cussing a storm and using a lot words that I had never heard before. It must have been some kind of Aussie bush talk. It sounded like a bulldozer was headed right at me.

The Aussie screamed out, "Here comes the bloody bull."

Trees were breaking, limbs were snapping, birds were flying out of the bushes, and this monster gray bull appeared out of the bushes from about 20 yards at a dead run right at me. I swear that bull grew while he was down there in the brush.

The trail was not wide enough for me and the bull. I quickly stepped off the trail and shouldered my club for the home run swing of a life time. Right when the bull came by me I planted a swing as hard as I could right on the end of the bull's nose to try and turn him toward the corral. He never even blinked or slowed down. My club went flying out of my hands on impact with his nose. It was the most helpless feeling that I had ever had in my life.

There went the earthmover of a bull over the hill crashing off into the darkness.

The Aussie came steaming up out of the woods with a heavy sweat covering his whole body. He screamed out, "Every darn tree limb down there had a thorn growing out of it!" He was cut up all over by thorns and briars. His face looked like you had smacked him with a wire brush. Blood was dripping off of his face all over the place. Now I had a mad, bleeding Aussie and a wild loose bull on my hands. The Aussie bellowed out, "The heck with the bull. I have got to get to St. Louis for the rodeo. It starts in two hours."

So there I was by myself with a mad 2500 lb bull running loose on my farm.

I went back to the house and strapped on my coon light for tracking the bull. It was easy tracking him in the dark. He left hoof prints in the ground 6" deep every time he took a step.

I finally tracked him to a brand new five-wire barbed fence, which was the best fence that I had on the farm at the time. The bull had gone right through it, and broke the top four wires! That was when reality set in. I was not going to be able to do anything with that bull. There was not a fence on my farm that would hold him. I silently backed out of the brush and left him alone.

I called the rodeo owner and told him what had happened with his bull. His reply was, "If you can not catch the bull, shoot him." I told you those rodeo guys were different.

The next morning I crept up to the area and could find no trace of the bull. I had to be at work in 30 minutes, so I left the area and hoped that he would stay around on the farm. That evening when I came home from work, I met my neighbor who I share a driveway with. Her eyes were as big as golf balls. She nervously exclaimed, "Do you know anybody who has a huge bull?"

I kind of sputtered out real softly, "I might know."

She answered back, "Well, it is up in my yard, and I'm getting the heck out of here."

Sure enough, when I crested the hill in my driveway, there was the bull standing between two fences right in front of her house. One fence was my five-strand barb wire perimeter-property-line fence and the other fence was a five-foot woven wire fence my neighbor lady had put up three feet away from mine because she didn't like my perimeter fence. It was on posts on six feet spacings. The two fences formed an alley chute that was 100 yards long. The alley was blocked perpendicular with a good cross fence that the bull was facing. The bull was so tight in between the two fences that he could not turn around.

I went home and saddled up my horse to try and work the bull backwards up the alley. I approached the front of the bull and started making lunges at his head with my horse. He backed up perfectly the whole 98 yards, then sprinted the full length back to where he was when we started. After three tries I finally got him backed all the way out of the alley.

The second he was free from the alley fence he charged me.

I kicked my horse in gear and side stepped him.

The monster bull took off running down my driveway towards my house. I was right on his heels with my horse, not sure what I would do if I caught up with him. He literally out ran my horse down to the house and jumped a hot wire fence in full stride.

I decided right then that I needed to call the local butcher. It was Saturday morning. The local locker owner told me that he was closed on Saturdays. He did give me his butcher's phone number and told me that if I could talk him into coming out, that was fine with him.

So I called the butcher and asked him if he could bring his truck out to butcher the bull. I also volunteered to help him.

He asked, "How big is this bull?"

I replied, "Oh, he is a fair size bull."

The butcher agreed to come out, but told me not to shoot him until he got there.

The bull was standing out in my yard about 80 feet from my house, pawing dirt. He had already broken off three beautiful 20 foot white pine trees in my front yard just by pushing on them. There was no way that I was going to take a chance of this bull getting away before the butcher got there. I loaded up my .243 deer rifle full with shells. I put in ear plugs, because I knew that I was going to have to shoot him from inside the house. I had never shot a high powered rifle from inside a house, but I knew it would be loud.

I quietly eased the screen door open and got a good solid rest on the door frame. I had a thought come to me as I sighted through my scope. What if I just stunned him and he took off running at my house. I would have to really act fast to get a second shot before he hit the house. Otherwise I would have two extra doors in my house - one door where he hit the house, the other door where he exited through the back side!

The bull was staring at me and pawing dirt with his head held low and his neck arched. He looked like at any minute he was going to charge me. I took good aim and squeezed the trigger. The bull went down and never moved.

The butcher arrived about 15 minutes later. He asked me where the bull was. I told him that he was around in front of the house.

I will never forget the butcher's face when he turned the corner of the house. With an astonished look he said, "My God, you didn't tell me you had an elephant out here. Look at the size of that beast!"

He backed his pickup truck up to the bull and hooked his winch cable to the bull. The winch cable ran through a steel A-frame that normally lifted the animal into the back of the truck. Once we had the head and entrails removed we turned the winch on. The winch started squealing and smoking just with the hind end of the bull lifted off the ground. Finally the winch just stopped followed by a heavy cloud of smoke coming off the winch motor.

Now we were really in a fix. The bull was half in and half out of the truck with no way to go up or down.

I ran to my garage and brought back a hi-lift hand jack along with a two ton ratchet come-a-long. One end of the come-a-long was hooked to the back feet of the bull, the other end was anchored in the steel stake eyelets of the pickup bed. The hi-lift jack was hooked to the bull's front section.

I started jacking up the front half and the butcher started ratcheting in the back half. All of a sudden the bull's weight shifted toward the butcher when I had the front section of the bull jacked up even with the bed of the pickup. The bull shot in the bed and busted out the back window of the butcher's boss's truck. At least he was in the truck after five hours of back breaking work. When the butcher got back to the cooler to hang him, they had to cut him in four pieces because he was too big to fit in the cooler.

The next week that locker owner told me, "Don't ever

call me about butchering a bull like that again. We are just not set up to handle animals like that!"

I don't know what he was so sore about! He only had to do one butchering job and got paid for the weight of two animals. Maybe it was the broken truck window.

We ended up with 1400 pounds of nice lean hamburger. I split the hamburger with the rodeo owner and came out with 700 pounds of free meat for my trouble.

I'm glad my rodeo stock grazing days are behind me, but I wouldn't trade the experience for anything. Unexpected situations are what build your character for handling other tough situations that may arise in your grazing operation.

Chapter 10
Losing a Lease Is Not the End of the World

I'm not going to go into detail about leases and contracts here. Both of those items were covered in detail in my previous book *No Risk Ranching*.

However, I am going to talk about losing leases. That's something I didn't write about in the first book and have since experienced.

We all have heard the horror stories from people who have worked hard on leased farms for years and then lose their lease for various reasons. This basically puts you out of business, at least on that particular farm. It can leave you feeling bitter, feeling cheated, questioning yourself, and wondering what you did wrong.

It is disappointing to work several years on a farm, get it fenced right, water established, etc., then lose the lease.

We have been leasing farms since 1999 and have lost two leases. I want to go over some of the things that we have changed concerning landowners and some early warning signs that we missed on the farms we lost leases on.

The First Lost Lease
The farm had 500 acres, with total grass pasture of 300 acres. The landowner commented that he had 500 acres of grass. I showed him a map of his farm that had the actual open areas measured off by the local NRCS office, but the landowner said the map was inaccurate. He insisted that he was renting 500 acres of grass, take it or leave it.

That was one warning sign. Another warning sign I ignored was that this farm was close to my home, so I thought I could still make it all right by running stockers on it.

The whole farm was in bad need of fence repair. I honestly don't know how the owner kept his cattle on the farm all those years. Well, thinking back now, I remember seeing them out quite a bit.

Another red flag was that the previous leasee had given up the lease because the landowner was very hard to get along with. But I ignored this red flag as well because I had grown up around this landowner, known him all my life, and just figured we could surely get along. I found out that knowing someone your whole life does not guarantee a peaceful working relationship when money is involved.

Another red flag that the landowner voiced was that he wanted his pastures mowed every summer to keep the place looking pretty. I agreed to buy the fuel and we would use his equipment to mow. The problem was that he mowed off my stockpiled forage so that his pastures looked like a dried up golf course. He simply could not grasp the importance of stockpiling forage. His reasoning was that you baled hay for the winter, not let the cows eat grass.

The one good thing I did with this landowner was I got a signed one year lease for $20 per acre, or $10,000 per year. I actually was paying $33 per acre of grass. The going rate in my area at the time was around $15 per acre.

In the contract I stated that all posts and wire that I put up were mine at the end of the lease.

I plowed ahead and worked day and night building fence until it would hold cattle. I made about $20,000 per year from custom grazing on the farm for three years. I had to run around 400 head per year to earn this. A lot of these were cows custom grazed at $15 per month. So it took me half the year just to make his lease payment.

If there had truly been 500 acres of grass to graze, I could have almost doubled my income. This farm had never been limed, and had never had any legumes seeded on it. The landowner's management had consisted of grazing the whole place continuously and putting down $5000 worth of nitrogen every spring. By the first of June he had to bushhog the entire farm because the fescue was so rank the cattle would not eat it.

I remember the first year I leased it, he told me that I had better put down a bunch of fertilizer or I would not have any grass to graze. After the second year of my management, the landowner commented that he had never seen his farm look so good. There was clover in all the pastures. I had fenced the hay fields into paddocks, and they were cranking out great forage. People driving by the farm were starting to take notice of the massive landscape change.

At the end of the third year, the landowner informed me that another fellow had offered him $30 per acre and he wanted to discontinue our lease.

Now that I had turned the farm around and fenced it to hold cattle, the landowner decided that he would take advantage of my work. At the time I had ten other farms that I was managing besides his. I was working very hard to maintain all of them and work full time in town besides.

I immediately went in the next two weekends and pulled all my posts and wire from his farm. It was a lot easier than I had thought. I simply wound the wire up on a spinning jenny and when the jenny was full, I would tie it securely in four spots and start the next spool. The posts were all fiberglass, and they came out easily.

The next weekend the new leasee came in to survey his newly acquired land and immediately went stomping into the landowner's home and asked where all the darn fence went. The landowner called me up and started chewing on me about removing all his fence.

I replied, "Look at your written lease."

It took them almost a full year to replace what I took out in two weekends.

The landowner had to pay for all the new fence out of

his pocket because the new leasee said that was the deal they had agreed on when he leased the farm. In other words he leased the farm with the understanding that it would have paddocks and perimeter fence that would hold cattle.

Today the farm is almost solely being used to make hay. The landowner no longer has clover in his pastures, just fescue hay fields. The worst part is that the landowner is losing all his fertility by selling the hay off the farm each year. In several years he will be left with broomsedge pastures.

I chalked the whole thing up to a learning experience. I am much better off today without it simply because I don't have to pay a $15,000 pasture lease every year. It has allowed more time to focus on my additional economical leases that are paying huge dividends.

Lessons learned:
1. Never sign a single year lease.
2. Never pay for grass that is not there.
3. Don't let your emotions of wanting the lease so bad override your common sense.
4. Do not attempt to work with negative people.
5. Trust your gut feelings.
6. Think things through entirely before proceeding.
7. Don't work with landowners who refuse to understand the importance of managed grazing.

The single thing that saved me on this lease was that I did have a written lease with the landowner's signature on it. If I had had a verbal contract, the posts and wire would have been history.

Looking back I can see how foolish I was in ignoring all the early warning signs. I was so eager to see what dramatic changes I could bring to this farm that I plowed ahead with my eyes closed. One thing I learned is that this grazing stuff is a lot more fun when you have landowners who appreciate what you are doing on their land.

It was a good learning experience. I'm glad we went through it, and we have learned some valuable lessons from it.

Second Lost Lease

The second lost lease was on 80 acres of unimproved pasture land. The land bordered another leased farm that we had. The 80 acres needed lots of work to graze cattle. It had very little fence, no water, and an absentee landowner.

I met with the landowner, who was very negative about everything right from the start. Being an optimist, I figured I could win him over once he saw what was happening on his place.

We signed a five-year lease, and I agreed to pay for building a pond if we could deduct the cost from the lease.

We secured the perimeter fence, built the pond, stocked it with fish, and fenced the livestock out of it.

Next we cleared the cedars from all the pastures to release the grass. We frost seeded clovers into the pastures, and cleaned up the brushy areas.

Next we subdivided the farm into paddocks, which allowed plenty of rest between each grazing since it was attached to our other leased farm.

Nothing we did ever brought one positive comment from the landowner.

The landowner was just not a very happy person about anything. Everybody in the world was out to get him, and he was convinced of that. After everything we did, he never got any friendlier as time went along.

The good part was that we did not have to be around him that often, simply because he seldom came out to the farm.

The bad part was that this lack of communication was the death sentence for the land lease. I'm not sure we could have ever changed his attitude. It was kind of a relief ending the lease after five years.

Lessons learned from this lost lease were good as well:

1. If a landowner is unfriendly, pessimistic, etc., right from the start, you will not change him. Thank him for his time and leave.

2. If we had gone the extra mile and kept in contact with him more, it may have helped some. I just don't like being around negative people. Life is too short to put yourself through misery that you can avoid by making smart decisions.

Chapter 11
Keeping Those Long Grazing Leases

There are good landowners. Today we have seven landowners who are all great to work with. They are just wonderful people, and they appreciate all the work we do on their farms.

We work very hard on keeping an open line of communication with them. We email them bi-weekly on farm events that are happening on their farm. These events include any kind of work that is being done on their farm - wildlife sightings, new forage species appearing, dung beetle sightings, newborn calves, lambs, pigs, goats, baby chickens, weather events, stockpiling pastures, or big fish swirls in the pond. We're always bouncing ideas off of them.

We call them personally and ask them to meet us at their farm and go over what we are doing. We ask them, "How do you think your farm looks?"

We also invite them out for dinner and go over the progress we are making on their farm.

We give them free samples of grass-finished meat. This meat is from our own steers and lambs that we graze on their farm. This allows the landowner to taste meat that is raised from the grass on their farm.

In 2007 we began to allow our landowners to purchase cattle to graze with our own cattle. We get to pick the cows with the right grass genetics, and they have entrusted me with that task. Their cows have their name on the cows' ear tags so that they can tell which ones are theirs. The landowner's newborn calves have their name on their ear tags as well. The ear tags are very visible and have the landowner's name written in permanent black ink on the face of the tag. This way the landowners can come out and look over their animals! It is very satisfying and enjoyable for a landowner to walk through the herd and look at their own cows and calves.

We feel like this will make them even more involved in their farm and the livestock. They are no longer landowners; they feel like they are partners. We are growing a friendship and future with our landowners.

We also feel like the landowners will be less likely to ever sell the farm when they have their own animals on their farm. Our landowners do not want to manage the livestock, so we are charging them a custom grazing fee for taking care of their livestock. By allowing the landowners to be cattle owners they get educated on all the details that make up a cattle business. We are very excited about this new endeavor. We love to see satisfied happy landowners.

In 2007 we had 40 things to do one Saturday morning, but instead we called one of our landowners and invited the whole family out to go horseback riding with us on their farm.

We had the horses saddled when they got there. Their daughter is a horse nut. She and her boyfriend had a wonderful time riding with us. Mrs. Landowner had a great time riding as well. She got to help me tag a newborn calf that day. She bragged to her husband when she got home how exciting it was to witness a new wet calf and hold it while I tagged it. Pretty much all day was consumed by taking time out of our schedule to take them riding. I don't think any of those 40 things that I had lined up to do that morning were as important as taking our landowners horseback riding!

We take out the video camera on their farms and video the progress that is taking place and send them a copy of it. This is a great way to show them the improvement while narrating what is going on with each process. With absentee landowners, this really puts them in touch with their farm.

We take lots of digital pictures and email them to the landowners as well. Digital cameras make it so easy, and it gives you a photo history of how you are doing from one year to the next. This is priceless information that you can use to make management decisions on that particular farm. When you are on the farm every day sometimes changes go unnoticed. These changes may be good or bad. You can visually see the changes when you compare one year to the next with photos.

When you start taking these pictures, drive a steel post that has a good landmark in the background. Fasten a sign on the post telling what month and year the picture was taken. Take the picture each year from the same spot with the land-mark in the background.

Some people may think that we go through a lot of hoops to satisfy our landowners. You would be right for thinking that. But if you lose your landowner, you have lost your leased farm and livelihood, so yes it is the most important thing you can do if you are in the leasing land business.

We have had several folks say, "You folks are sure lucky you have all those cheap land leases or otherwise you could not graze all those cows."

Folks, luck has absolutely nothing to do with it. Fostering landowner relationships, determination, honesty, hard work, good attitude, planning, reading, and constantly learning new concepts will put you on the right path toward building a profitable grazing enterprise.

Chapter 12
Reducing Fossil Fuel Consumption

Before we go to the next section, I have a question for you.

How many of you think fossil fuel is going to get cheaper?

You would have to be a huge optimist to believe that fossil fuel is going to go any other direction than up. As I write this, oil is at $100 per barrel. We still have time to get our operations fine tuned to drastically reduce the consumption of fossil fuel.

I don't see any cheap fuel sources on the immediate horizon that will replace what we have today. All of the alternative fuel sources that we are starting to mass produce are expensive to make, and I'm not sure any of the alternative fuels are sustainable.

So what are we to do for fuel?

I'm suggesting that we will always have some need for fuel, but the key word here is "some." Most livestock operations today are heavily dependent on fossil fuel to turn out a product from their operation that they already have in place.

Most people are resistant to change because it makes them extremely uncomfortable. Those producers who do not take the initiative to change right now will not be having a very fun time of it when fuel prices double.

I see all these cattle producers out mowing, raking, baling, hauling, feeding, reseeding, spraying, etc. and all require fossil fuel.

What if fuel goes to five to eight dollars a gallon? Will those operations still be able to do all those functions and sell a dollar per pound calves and stay in business?

What if calf prices go to $.60 per pound? You will hear a lot of livestock producers moaning and groaning, including me. With our low cost of gain due to low overhead, no grain, no outside costly inputs, cheap land leases, very low fossil fuel usage, mob grazing, mimicking nature, we will still be showing a profit at $.60 per pound commodity priced calves.

I am not trying to brag, just stating the items that you need to be focused on to get in the life boat with us. Hey folks, this is one fun ride with our current price of $1.25 per pound calves at the local auction markets. We can do a lot of things wrong and still make money because of the huge cushion that we have built into our grazing system.

It makes no sense to fire up a 150 hp tractor to bushhog off a field when livestock can do it for free and fertilize it at the same time. In the next 10 to 20 years we could easily see six to eight dollar per gallon fuel and here's why. Drive out through any large city and look at the eight lanes of solid cars going every which way. It is truly amazing that we can keep the big ocean tankers coming fast enough to refill our gas tanks.

Look at China and India. Their economies and standard of living are increasing very quickly. What happens to fuel prices when everybody in those countries begins driving cars? China just built the largest interstate system in the world. They have 28,000 new car owners every month being added to their fuel consumption.

World fuel consumption has not peaked, but fuel stock piles have peaked according to the oil companies. There are no more huge undiscovered oil reserves out there. Oil is at $100 a barrel. It doesn't seem like that long ago that a barrel of oil was under $10.

The purpose of this chapter is not to scare you, but to wake you up to the real fuel concerns on the horizon that will have major effects on most livestock operations.

Fossil fuel is a consumptive item on your operation. Once you use it, it never comes back. It does not regenerate itself.

Livestock are the opposite. They can regenerate themselves every year with only a minimum of fossil fuel required. Heck, they are the ultimate machine. They can run on solar energy when we allow them to. What's better than that? A machine that reproduces itself and runs on sunlight.

It still amazes me every day to see people operate in the cattle business and put absolutely no value on their grass.

Just two weeks ago, September 15th, 2007 (peak grass stockpiling time) we had two neighbors mow off their entire pasture with a rotary mower. Those pastures had 10" of some the prettiest grass you ever saw. One pasture covered 80 acres, the other one around 60 acres. They both mowed it off like it was their front yard, about 2" stubble was left. This was in the middle of a major drought!

This morning, one month later, those same two cattlemen had their cows locked up in a lot feeding them big round bales of hay. They both have put themselves into a very long costly winter of feeding hay every single day until spring - six months of feeding hay until April arrives. Doesn't that sound like fun to you?

Can you imagine how many days of good forage they had out there if they had been willing to strip graze those areas? Thousands of dollars of good grass was wasted because they wanted their pastures to have the nice neat mowed and manicured look. What a price to pay for a mowed lawn appearance.

Look at the fossil fuel that was wasted by mowing off perfectly good grass that the cows would have loved to graze. Both of those fields were heavily fertilized this year as well. So they paid to put down expensive fossil fuel consumptive fertilizer to grow the grass, then mowed it off on the ground and let it lay in September.

This is the prime period that we generally grow our winter stockpiled grass for the whole dormant season. Those two fellows must have very deep pockets and put no value on their grass at all. It is strictly a hobby and a tax write off for them. I think I could find a better hobby than getting up every morning all winter in the freezing weather, trying to get the tractor started to feed hay. I would hate to see their hay bill at the end of the winter season.

Folks, we are in the best business in the world. Sell your heavy metal, buy livestock and a wheelbarrow. You will need the wheelbarrow to pack all your money to the bank!

Be careful who you follow in making decisions on who to buy cattle from. There are some real fossil fuel consuming cows out there. What appears on the surface as a quality beef herd initially may have some very deep flaws hidden when you dig deeper.

Let me explain. I went by a fellow's place where he had some bred heifers running on pasture. The pasture was junk, no rotation or rest for the grass, eaten right off flush with the ground. The heifers were all slick, shiny and fat as pigs.

My first response after seeing them was, "Wow, what a beautiful set of heifers considering they have no pasture. Those heifers must really have some good genetics to stay that fat and slick with minimal forage available."

The rest of the story was filled in later when I visited the guy early one morning. When I drove up to his place, he was on a tractor auguring out a mixed ration of grain into feed bunks for the heifers. Heck, he was feeding them like pigs! No wonder they were fat and slick. The heifers were fat enough to butcher, not to put in your cow herd to raise calves.

The last thing you want to do is to feed them grain if you are thinking of putting heifers in your cow herd. This builds fat in their udder and causes major problems with milking ability when they calve. Once that fat is built in the udder while heifers are developing, they are ruined for the rest of their lives. So my initial response to what looked like awesome grass genetic heifers was all a lie, hidden by grain feeding.

The guy would have been money ahead to have sold

them to a feedlot instead of putting them in his cow herd. Basically what the guy had done by feeding those heifers was bury the future of his cattle operation. Now, the future of his whole herd is dependent on fossil fuel, not solar energy.

I know from experience. I did it when I first got into the cattle business. I was inexperienced enough to not know any better. I just assumed that because everybody else in the neighborhood was feeding grain to their livestock, that was what you did. I am just lucky that I smartened up quick enough to limit the financial damage to my operation.

I still remember the first load of bulk feed I picked up at the feed store. It was a recipe that would make me a lot of money, I was told. My cattle would stay healthy and fat, which guaranteed me lots of calves with mother cows that bred back. That first load cost me $300 a ton. I remember how poor I felt when I drove out of that feed store, but heck, I was going to be rich as soon as I fed those cows that purchased grain mix! This included grinding, mixing, vitamin packs, molasses, etc. Corn prices at that time was $2 per bushel, and 400 pound calves were selling for $.60 per pound. Heck, I was feeding the equivalent of one calf a week just to keep the cows going through the winter period.

I was hand feeding them once a day, just a couple of pounds per head. Do you know how long a ton of feed will last a group of beef cows in the winter? Not very darn long! You can not feed supplement your brood cows and expect to show a penny of profit. Your brood cows have to perform on pasture alone or you don't have a profitable grazing operation. I do not want you to go through the misery that I mistakenly did.

If you're going to own cows, select genetics that do not need the almighty fossil fuel machine, and your bank account will thank you. You also will have a very bright future waiting for you in your grazing operation.

Multi-species Grazing

Chapter 13
Diversify Your Landscape: Multi-species Grazing

It has been a great experience learning multi-species grazing. I recommend it 100% to anybody who is interested in making a true positive difference in your grazing business.

I dragged my feet for years, using a whole list of excuses of why multi-species grazing would not work. The extra fencing required was a major excuse for not grazing sheep. I mistakenly thought that I would have to build another special handling facility for the sheep or goats. I had read a lot of articles about livestock guardian dogs and they sounded like a huge expense coupled with very short life spans. The constant worming of sheep and goats did not appeal to me very much either.

I was always told that if you owned sheep you had to have a Border Collie to herd the sheep when you needed to catch or move them. Good Border Collie dogs are very expensive and would be another mouth to feed.

Of course, we have all read about how sheep destroy pastures by eating the grass off flush with the ground. Everybody I've ever known who raised sheep always had some kind of shelter for them and usually lambed them in a barn. Building a shelter and lambing in a barn did not appeal to me at all. Most of the sheep operations that I had been around were very labor intensive.

Other sheep folks told me that you had to feed the ewes grain in the winter to flush them so that they would have twins, and they also needed alfalfa hay in the winter. Feeding alfalfa hay and grain all winter sounded like a money drain to me. Trimming hooves of sheep and goats did not sound fun either. That pretty much sums up my list of excuses why we never grazed sheep or goats.

Now, I am going to go back through each excuse and explain how we worked through each one in our operation once we decided to give multi-species grazing a try.

1. Extra fencing:

I had figured that if you grazed sheep you had to have a woven wire fence to keep them in. Woven wire is very costly to put up. It does make a nice fence, but it will break your bank account when fencing a large area. All of our sheep fence that we built was erected using multiple hi-tensile electric fence wires. We use four strands for interior fences and five strands for perimeter fencing. This type fencing is fast, much cheaper and keeps the sheep in.

2. Building an extra handling facility:

We did not build a special handling facility for the sheep or goats. We used our existing cattle handling facility with some slight modifications to handle sheep and goats. We used a roll of woven wire and lined the inside of our cattle corral with it. The woven wire has plenty of strength with the stout cattle panels backing it up for support. The actual alley that leads into the head chute is where we do our sorting of the sheep. We built a portable alley and guillotine gate that allows us to sort the sheep one at a time. It works great and was built out of scrap lumber.

3. Costly livestock guardian dogs:

Our oldest guardian dog just turned six years old and she still acts like a puppy when you pet her. She shows no sign of aging, and is always out working at keeping the predators at bay. So my worry of them having a short life span really has not materialized.

We gave $200 for our first dog, $125 for two others. We had one very good male dog given to us, so dog purchasing cost has not been an issue either.

The guardian dogs do not eat very much dog feed in the summer, but probably double their consumption in the cold winter months. Our dogs get lots of meat scraps from our local butcher. They stay in good shape eating meat trimmings and fat scraps.

4. Worming sheep:

We are focused 100% on building a parasite-resistant flock, so worming is not an issue. We have never wormed a sheep. If we have a sheep that shows worms, we sell it.

5. Sheep herding dog:

A herding dog is not required to gather sheep or move them. Our sheep are trained to move by calling to them. I will admit there have been a couple times that a well trained Border Collie would have come in handy to move them when they did not want to move.

6. Sheep destroy pasture:

Any type of livestock will destroy a pasture if you leave them on it long enough and do not let the pasture recover fully before grazing it again. We rotate with a 14 -18 week rest period between grazings.

7. Sheep need shelter:

Our sheep do not need shelter, and they perform just fine without it. We lamb on the pasture, not in a barn. Sheep have such a thick coat in the winter that it allows them to stay warm no matter what the temperature is.

8. High quality hay and feed required:

Our sheep have never had a bite of alfalfa hay. Our sheep are wintered on stockpiled forage. What little hay we do feed in ice storms is strictly grass hay. We do not feed grain to flush them either. Feeding grain is a total waste of money.

9. Hoof trimming of sheep:

We have never trimmed a hoof on a sheep. Our hair sheep do not require hoof trimming because we feed no grain. Feeding grain will grow excessive hoof material.

Summarizing, all of the concerns that we previously had really turned out to be nothing to worry about after all. Every concern was worked out with no problems or excessive expenditure of capital. You just have to get over the hump of each concern by concentrating on solving it sustainably.

People have asked us what we do differently with the management of sheep and cattle.

Sheep are not cattle, but we treat them about the same except for the guard dog protection. The only changes we made to the cattle fencing was to put in additional wires to keep the sheep in.

Keep it simple and brainstorm solutions to remove any potential roadblocks. I am convinced we sometimes make things much too complicated. I know that I am guilty of that sometimes.

When you mix two or more different species together neat things start to happen. What was once nuisance plants, weeds, brush, etc., become food for your animals. The cattle eat the grass and legumes. Sheep go after the broadleaf weeds, blackberry and some grass. It really is amazing to see what sheep impact does to a field of weeds. A once prominent field of weeds is turned into weed stems, a truly wonderful sight. Weeds are turned into quality meat. It just does not get any better than that. Goats go after the more woody plants. (More on goats later.)

We do not custom graze sheep. We own our own flock. If you had access to some sheep, custom grazing would be a good way to learn the basics of grazing sheep. It would give you some income and weed control.

Sheep do not need quality pasture to perform. If you have some brushy ground, the sheep will eat everything they can reach. We use them to help clean up our leased farms.

Several years ago we had an ice storm come through that left 3"of solid ice and then a 20" snow covered the ice. This is a grazier's nightmare. The sheep spent almost all of their time down in the woods foraging. The woods area did not have the ice layer like the open pastures had.

The sheep would use their front foot to paw down through the snow and expose the bare ground. They cleaned up all the leaves and acorns in the woods for the six weeks the snow was on the ground. The whole woods looked like a bunch of hogs had been rooting around in it. They truly are aggressive foragers.

We have tried several different grazing systems with the cattle and sheep.

Sheep and cattle together. We have one farm where we graze sheep and cattle together. The cattle are rotated daily and the sheep graze where they want. They mainly go after the broadleaf weeds that the cattle ignore. The loosely stocked goats go wherever they want to on the farm.

The cattle act as dead-end hosts for the sheep parasites. The sheep soak up the cattle parasites and act as a dead-end host for the cattle parasites. The two species complement each other in keeping down the parasites and improving the forages in the pasture.

Yearling calves, then lactating ewes, followed by dry cows. Another rotation we have used with cattle and sheep was yearling calves grazed first, followed by the lactating ewe flock, then dry cows cleaned up what was left. By the time the dry cow herd exited the paddock there was not much left. This is not a problem as long as you let it fully re-grow before repeating the process.

When we used this process, the stocker calves only ate the very best of the plant tips because they were moved every day to a fresh strip of grass. Once the calves left the paddock, the lactating ewe flock was turned in to graze. The ewe flock had access to the whole paddock that was strip grazed by the stockers. They cleaned up the middle parts of the plants that the stockers left and most of the weed leaves that they could reach.

After the ewes were moved to the next paddock, the dry cow herd would be brought into the paddock to clean up what was left. With this method we were getting about 90% utilization of the paddock, which is not a problem as long as you allow the plant to fully re-grow before grazing it again.

Alternating sheep and cattle rotations. The grazing system that we use with our large sheep flock is the one we use the most because it requires so little labor. Our sheep flock is moved every seven days to a fresh rested paddock. These paddocks range from 5-10 acres each and have four strands of hi-tensile electric fence on all four sides of each paddock.

The cattle are rotated through the sheep paddocks at least twice during the growing season. One revision we plan on making to this system is to put about five dry cows with our 300 sheep. This way the cattle can soak up the parasites immediately and it will also give us the diversity of two species grazing together. The reason for only five cows is because with the five acre sheep paddocks, the sheep flock would run out of grass before seven days were up if we had a larger number of cows grazing with the sheep.

Some people have asked which grazing system gives us the best parasite control?

As long as you give a sheep-grazed paddock exposure to cattle, the parasite load is going to be lower. Then add in a long rest period before the sheep come back to that paddock and you have a grazing system that is structured toward low parasite levels.

I believe in a humid environment that you will always have some parasites in your pastures. The trick is to develop a sheep that can perform perfectly with a low parasite load. This low parasite load helps the sheep develop a strong immune system to protect itself. If you are constantly worming the sheep, they will never develop a strong immune system. All you're doing is placing a crutch under them by worming.

It is so nice to keep the brush and weeds in check with animals rather than constantly fighting them with herbicides you constantly have to repeat every year. The worst part is that you're putting poison down on your pastures and killing a lot of beneficial forbs, microbes and insects in the process.

It is also very costly to apply herbicides. I hate the smell of them and the mist landing on my clothing and skin. Again, we would rather work with nature than against her. It's more sustainable for sure. I never looked at weeds and brush as a money maker before. Now I do.

Another advantage of grazing sheep or goats with cattle is the added predator protection the cattle give you. We lamb right out in the pasture on the farm where the two species are grazed together. It is so darn natural. There is no better sight than to watch the two species grazing peacefully intermixed throughout the paddock.

One goal we have is to actually get our own herd of cattle bonded onto sheep so that they always move as one herd. If we can do this, the need for a guard dog could be eliminated. The key is to keep them in small enough paddocks so that the stocking density is high enough to keep predators at bay.

We feel like we are spreading out our exposure to price cycles by having the multi-species as well. When cattle prices are high, sheep prices tend to be low and the reverse is often true. Sheep are another species that can help make a living for you on your farm by basically using the same infrastructure that you have in place for your cattle. You may want to think about giving hair sheep a look. You won't have all your eggs in one basket, which makes your income more stable.

You may be thinking I forgot to tell about where the pastured hogs fit into the multi-species grazing. I have a whole chapter later on explaining the pastured pigs and how they fit into the grazing operation.

Chapter 14
Building Fence for Sheep and Goats

To keep with the paddock rotation when adding sheep, we have added three wires to our existing single wire paddock divisions. If you want to hold goats and do paddock rotation it may take a total of five wires. All the posts were already in place. We just added the wire to them.

We have cut the cost down by drilling 1/8" holes and fastening the hi-tensile wire to the post with 12-gauge galvanized flexible wire. Cotter keys that you buy from fence manufacturing companies run into a lot of money when you start adding multiple wires to every post.

One tip you want to remember when fastening these wires to the posts is be careful not to snug the wire up tight against the post. If you snug it right up to the post, you remove the flexibility of the hi-tensile wire to move when something hits the fence. It may even break at the point against the post. Give each wire three twists and leave a quarter inch space for wire to move. It also makes ratcheting up the wire a lot more efficient if you have this space for the wire to give.

We had to add some posts in areas where the ground was uneven because you don't want your bottom wire to touch the ground. Our wire spacing for the sheep fence consists of 6" off the ground for first wire, followed by 6" for the second wire, then another 6" for the third wire, then 12" up for the final fourth wire. This gives us a 30" top wire to stop cattle.

The second wire from the ground is hot and the top wire is hot. The first wire and third wire are hooked up to the grounding system. By making the bottom wire ground, this helps keep the grass from drawing volts out of your fence. In prolonged wet periods, that bottom wire will see a lot of moisture, which hammers your voltage output.

We drive ground rods throughout the paddocks spaced about a quarter mile apart. These rods are hooked onto the ground wires at that point. This way when an animal touches the wire the animal is instantly grounded and the ground path does not have to travel back to the charger where your primary fence ground system is. This works excellently in dry conditions also. If your animal is not grounded when they touch your fence wire, you can have the hottest fence in the world and it will not keep your animals in. At every corner all wires are crimped together to their respective wire, this way your entire grazing system is tied together.

To carry the electric power across any gate areas, we use the following system: Let's say you have a 10-foot gate crossing in your power fence. Use a spade and a dig a trench 8-10" deep x 3" wide across the gate opening from one corner post to the other corner post. Lay a three-quarter-inch diameter black polyethylene pipe down in the trench. Each end of the three-quarter-inch polyethylene pipe should be angled up to the edge of each corner post and left sticking up out of the ground 3".

Tamp the dirt back in the trench covering your pipe. Cut a piece of black insulated underground wire that will run the length of the gate. Make sure when you cut your wire length that you leave enough extra to secure to the inside of each corner post and crimp onto each respective hot wire.

By running your insulated wire inside the pipe this protects it from rocks or whatever it may come in contact with.

At first I did not use the pipe conduit and paid for it later with shorts in my buried wire. It is a very sick feeling to hear your buried insulated wire popping under your gate opening. This buried insulated wire is expensive. You do not want to have to replace it because of a short in your wire. The gate area sees a lot of intense pressure, which causes rocks to cut through your insulated buried wire. I've never had a problem since switching to the polyethylene pipe as a conduit for the insulated wire.

One of the problems you encounter when running multi-wire paddocks is installing cost effective gates that will keep in sheep, goats and guard dogs. We have developed a gate that works very well, is super cheap, and is hot.

Remember the old wire gaps that were made with 4-5 strands of barbed wire? That is exactly how our hot-wire gates are made. Bear with me and I will try and explain exactly how we do it.

You begin by taking a four-foot length, 1" diameter fiberglass rod and place it at one corner of where you want your gate to fasten to. This will be the side that you open.

Take a piece of hi-tensile wire and place it at the bottom of the existing corner post by wrapping one end around the corner. Crimp it with a wire crimp sleeve. Now, you have this small circle of wire crimped for your bottom gate post to fit into.

Drill a hole into the existing corner post 2" off the ground and wire your hi-tensile circle to it. This holds your bottom gate catch off the ground and makes it easy for you to place your gate post bottom end into when securing. For the top of the gate post repeat the same process as the bottom end hi-tensile catch wire. Place the gate post in the top and bottom hi-tensile catch rings.

Take a tape measure and mark out the location of your wires on the fiberglass gate post. These measurements should be the same as your regular hi-tensile multi-wire fence. Take a quarter-inch drill bit and drill a single hole at each marked location through the fiberglass rod. Make sure they are all drilled in a straight line. Each hole should be directly above the one below it.

Next take and tie a piece of half-inch white poly tape to the bottom strand of the existing hi-tensile wire at the gate corner. Thread this piece through that pre-drilled hole in the bottom hole of your gate posts. Pull this piece of poly tape tight and wrap it around the gate post once before tying a knot in it.

By wrapping the poly tape around the post, this takes the pressure off of your knot and gives it more surface area to pull against.

Next, do the top piece of poly tape the same way as you did the bottom strand. Now, you have the gate secured and tight with a bottom and top poly wire.

Repeat the same process for the remaining additional strands.

I love these gates because you can keep them so tight.

This is critical because a tight gate will give off a shock better than a sagging wire gate. These poly tape gates are so visible. I've never had an animal even try to go through one. Where you latch the gate over the top of the post, this area gives you a good handle to hold onto when opening the gate. Remember these gates are hot when you handle them. I've never been shocked yet, simply because I have such control of them by having all poly tape secured to the post.

Another advantage of having the gates tight is that the wind will not whip your poly tape around. Wind can really work on the poly tape over long periods of time if it is fastened in place loosely. The plastic threads start to separate, then it loses all of its strength and breaks.

I like to make my gates 12 to 16 feet wide. The 16 foot works better for large groups of cattle. I have made some 20 feet wide, which work fine.

You simply cannot build a cheaper multi-wire gate that I know of that keeps any class of livestock where you want them. We can buy a 1320' roll of white poly tape for $40 - that comes to 3 cents per foot. A 12' five-strand gate would equal 60 feet of poly tape. We will throw in another 5 feet for each gate for wrapping around posts and knots. 65 feet x 3 cents equals $1.95 per 12' gate. Kick in 4 bucks for the 1" fiberglass posts and you have a total cost of $5.95 cents per gate.

Keep in mind that with paddocks you will have lots of gates. This is a huge cost savings when you start trying to compare the costs of this multi-poly tape gate against the purchased spring gates that are a pain to handle. You have five spring gate wires that you've got to keep track of. I can not even comprehend tackling that nightmare. But let's assume you want to, so let's calculate the costs.

The spring gate and handle package will cost 12 dollars

each. $60 per gate will drain your pocket book pretty fast. The worst thing is the animals run through them and stretch them out. Not only do you have an expensive gate, you have an ineffective gate to boot. I have several spring gates that I purchased before I designed the poly tape gate. Almost every one of them has been sprung by animals running through them.

With multi-tape gates, the tape does not get tangled like spring gates do. One handle and your gate is open. When you open the gate, have a one-inch hole in the ground back next to the fence that you can set the bottom of the post down in to keep it off the ground. You may want to leave the gate open after grazing that paddock. This keeps your poly tape from popping on the ground.

Another added bonus with having electrified poly tape gates is that it alerts you if you hear no popping sound when opening the gate. Your fence has quit or is grounded some-where. This has saved me numerous times. I have done it enough now that I can immediately tell if my fence is at full power or not, just by listening to how loud the popping noise is. It doesn't sound very scientific but it sure works.

Chapter 15
Those Wonderful St. Croix

One of the main reasons we got sheep was to control the weeds, multiflora rose bushes and thorn tree sprouts. They have flat hammered all of these species.

It is so rewarding to watch sheep pick the leaves off of a worthless weed. They eat the leaves like lollipops. You can definitely tell where we have had sheep after one year by the absence of these species that the cattle refuse to eat. We have several fields that we call the brush paddocks because of all the rank brush. It is very steep terrain with lots of brambles of every kind. The sheep do better on this paddock than the cattle do. They handle steep terrain very well.

Our landowner was kidding me the other day by saying, "We may have to rename those brush paddocks something else." The sheep flock has made a major difference on how those paddocks look today.

Most of the blackberry bushes have been killed by the sheep defoliating the plants. Depending on how much area you give the sheep each day, they will greatly affect the kind of grazing results you have. If you put them on a small area, they become very non selective. Everything is consumed.

We started with a flock of parasite resistant hair sheep. The hair sheep is strictly a meat animal; they have no wool. This has become a huge advantage because you don't have to pay someone to shear them.

To have a sheep sheared in our area costs more than the price you receive for the wool. It doesn't take a rocket scientist to figure out that you don't want wool on a sheep unless you like shearing them yourself. There are a lot more fun things to do than shear sheep. Your back will thank you for it. The hair sheep shed off their winter coat in the spring just like the cattle do.

It takes 20% of what a sheep consumes to grow wool that is worthless. This 20% could be growing additional meat that you can sell. You also do not have to dock their tails when they are born. They have no wool on them to catch manure and build a refuge for maggots.

We were lucky enough to find a gentleman who had a small flock of St. Croix hair sheep that he had never wormed. He agreed to sell us seven bred ewes to get started.

Initially when he started with the flock, he lost several to parasites, but he stuck to his guns and never wormed them and never lost any more.

We are excited about our endeavor with these wonderful animals. To give you some background on the St. Croix, they originated from the tropical island of St. Croix where they adapted to the tropical weather. The ones that didn't adapt died. They came over to the United States in the early part of the 20th century.

The adult ewes will weigh from 110 to 130 lbs. Mature rams will weigh around 130-150 lbs. Spring born lambs will average around 6-9 lbs depending on whether there were twins and the age and size of the mother. By the following winter the spring born lambs will average around 60-80 lbs depending on the growing season for that year.

The travesty is that most people who own St. Croix are worming them and have destroyed their natural parasite resistance. Whenever you worm an animal, you are propping an artificial crutch under them to survive in their natural environment. This weakens their immune system and pretty soon you have to worm them or they die.

The biggest predator in the sheep business in Missouri is not the coyote or stray dogs, it's parasites. Parasites kill far more sheep and lambs than all the other predators combined. The only way to build a sustainable, profitable sheep flock is to build a parasite resistant one.

First of all, the wormers on the market are becoming less and less effective every year they're used. The drug companies come out with a stronger wormer and the parasites soon adapt. There's nothing sustainable about that.

I had a fellow sheep friend tell me that there are producers giving sheep wormer orally that is labeled to give under the skin, which is the only way they get any kind of control. Producers are having to worm lambs monthly to keep them alive. This parasite problem is what keeps a lot of people from grazing sheep. They don't want the hassle of monthly worming, and I don't blame them. The cost of the wormer is high and the labor costs, even if you do it yourself, is astronomical.

Back to the St. Croix. Our goal is to build a flock that needs no wormer, no lambing assistance, no feed, no hay, no shelter. They must have extreme flocking instinct, lamb on pasture, get no shots of any kind, wean their own lambs, shed off every spring, lamb every year, require no hoof trimming, and have good mothering instincts.

We have a small flock that is doing all the above right now. But to get our numbers up to where we can make some money we feel like 500 ewes is a good goal.

We called all over the United States checking on more sources for St. Croix ewe lambs. I will never forget one guy asked $500 for his ewe lambs. I would be afraid to let those things out of my sight if I gave that kind of money for a lamb.

The fellow who we bought the St. Croix from encouraged us to buy any kind of hair sheep ewes that were reasonably priced and breed them to a St. Croix ram. Within several years we would have a major influence of St. Croix in our flock and be well on our way to building up numbers. So we bought 200 head of Barbados hair sheep from Texas to breed to our St Croix rams.

The hope is that the St.Croix rams will pass their

parasite resistance to their offspring out of the Barbados ewes. Each year we plan on keeping the ewe lambs out of that cross and breeding them right back to the St. Croix rams. Each year the offspring should be more parasite resistant. That is our plan.

In the mean time we still have our original pure St. Croix ewes pumping out twin lambs each year. If our Barbados/ St. Croix cross plan fails, at least we have the pure parasite resistant St. Croix flock to fall back on. It just will take us longer to build our numbers.

At the present we are in our fourth spring of Barbados/ St.Croix cross lambing. I can hardly wait each day to go out and see what the new lambs look like. It is quite a sight to see. Most of the Barbados ewes are brown on the back and have black bellies. Some are pure black, some are pure brown, some are tan, some pure white. It looks funny seeing a black Barbados ewe with two white lambs running along beside her. Some of the lambs are coming out black and white, some have black bodies with a white head and white tails. Some have brown spots over their dominant white body. It looks like somebody went crazy with a box of crayons. Eighty percent are solid white.

I've never had sheep before in my life, but I love them. The only management we did last year was removing the ram lambs before they were old enough to breed. At four months there are reported instances of them breeding. St. Croix are non-seasonal breeders. If you want to push them you can get two lamb crops in one year. Our sheep have to make it on forage alone. I've never experienced any kind of livestock that required basically no care, just predator protection and salt.

We are lambing in sync with Mother Nature - May 1st - right when the grass and clovers are starting to take off. For the life of me I can not figure out why people torture themselves and their sheep by lambing in the middle of winter. There is nothing natural about this. There is no succulent forage to feed the ewe so that the lamb can have some milk produced by nutritious grass.

When lambing in the winter you have to get the ewe up in the barn to prevent the lambs from freezing to death. It's not my idea of fun or sustainability. Why go through all that work and extra feed costs when nature does it for you free?

The flocking instinct of the St. Croix is wonderful. They always are in a tight circle when grazing. This is priceless when it comes to predator protection. It is a lot easier for a predator to pick off a sheep if they are by themselves. It is also easier for a guard dog to protect them if they are in a tight flock. I've never lost anything to a predator yet.

I've also never experienced any species of anything that acts like a family unit as the St. Croix. If a lamb is bleating, every sheep in the flock is over there checking out what the problem may be. When these sheep lamb, they are very protective. They even charge the guard dog to keep it away from their lambs. I've actually seen them hit the dog hard enough to roll them on their side.

The Barbados ewes have surprised me in their superb mothering instinct. These ewes will flat protect their lambs. I went out to feed the guard dog one night during lambing season. I called to her and she would not come to me. She had never done that before, which made me think at first that she was sick. Then she started acting real timid and cowering low to the ground. To come to me she had to pass right by some new lambs that had the ewes with them. She circled out around the ewes and they still came after her at full speed. I was impressed to say the least.

We run sheep and cattle together on the same farm and do not separate the sheep from the cattle during lambing season. The cows do not bother the ewes. Anything that approaches a ewe with a new lamb gets a warning to back off. The ewe will face the cow and stomp it's front food, warning the cow to back off.

They are a little bit like a cow when they first calve. They stay back away from the main flock for a day or two until the lambs are stronger.

Lambing season is the most entertaining time of all. Acrobats and gymnasts could learn a few moves from week-old lambs. I could watch them buck, twist, kick, rear, and run for hours. I've never watched a movie that was as entertaining as watching new lambs play.

They start nibbling grass as soon as they are three days old. They only take the tips off. It's quite an amazing sight. They come out tiny, but within a week of nursing the ewe they really put on the weight.

We do not interfere with the ewes during lambing season. They must be able to deliver the lambs on their own. Our experience has been that the more you mess with the ewes during lambing, the more difficulties you have. Let them work it out. The majority of them do fine without human intervention. This is the real beauty of lambing on pasture in the spring or early fall.

When you get out there and start playing mother, you upset the ewes and may end up with some abandoned lambs because you spooked a young mother away from her newborns. You will lose a few lambs, but this goes along with the sheep business.

When it gets close to lambing time, we plan out the lambing paddock ahead of time. We want the paddock large enough so that we do not have to move them for three weeks. All lambing is done within this time period. I wish cattle could do that.

The reason you do not want to move them is that the newborns and new mothers do not move very well. Anything that is left behind is extremely vulnerable to predators. We have moved flocks during lambing because we ran out of forage on the paddock that they were in. In these instances, you just have to be patient and let the sheep move at their own pace into the new paddock.

The ewes that lamb on their own without intervention are the kind of ewes that you want to build your flock with anyway. If you have to baby them, you can expect more of the same the next year. Don't get into this mothering mode. Keep your distance from the ewes with new lambs. They will do fine without you. The last thing you want to do is build a flock of pampered sheep.

Chapter 16
Developing Your Own Parasite-Resistant Flock

If you want to start your own flock of parasite resistant sheep there are several things you must develop.

First thing is to control your emotions. You must be tough enough to not give up when things are going against you. You will lose sheep to parasites if you are in a moderate rainfall area. You will be tempted to throw your hands up and start worming when you start losing sheep.

Worm the ones that look wormy and sell them. Don't worm the entire flock. You will destroy any chance of developing a parasite resistant flock.

Select sheep from your area. They will be better adapted to your environment and forage.

When we turned out the Texas Barbados Blackbellies on the farm it was lush with lespedeza, and they just walked around looking for some Texas brush to nibble on. The Texas sheep were from an area that received 10 inches of average rainfall per year. In Missouri they say our average rainfall is 38-40 inches, although we have not seen those amounts for several years. These sheep had never seen such lush forage, and didn't know exactly what to do with it. It was really hard for the sheep to adjust, and the parasites wasted no time jumping on their host animals. Our seasoned St. Croix never missed a beat during the rainy hot weather. The Texas sheep suffered.

Go on the internet and place ads on some of the sheep web sites stating that you're looking for some pasture-raised hair sheep. We bought 25 Barbados ewes this way in Missouri.

Don't get in a hurry and pay more for them than your wallet can afford. Start small, buy a dozen or so, and learn from there. It's a lot easier on your pocket book to lose four out of 12 than to lose 50 out of 200 like we did.

Try to find someone who is also building a parasite resistant flock. They may sell you a few ewe lambs. Stay away from producers who are babying their flock. You don't need to inherit somebody else's pampered pets.

Talk with them about their sheep and you will learn quickly what kind of care they receive. If they are feeding grain and alfalfa hay, stay away from them. You can not afford to feed either of those items and make any money in the sheep business.

Take pictures of your sheep for your records and date them so that you can track your progress visually. I can hardly believe the difference in one year of how much our remaining Barbados have improved in appearance. I was looking back at pictures of them when we first got them and comparing them to this spring. They didn't look like the same sheep. Their rumen has finally adjusted to our forage. The parasite prone ones have died or been sold, leaving us with a pretty good set of ewes.

We took manure samples of each St. Croix ram and had the vet measure the parasite levels in each. This measuring method is done 10 times from the same manure sample and an average is given. We were pleasantly surprised that most of the rams had less than .05 % levels of parasite eggs. One ram had 0%, not one egg. Another ram measured .03%.

This was a very pleasant surprise. These two rams are also the first to shed off 100% of their hair in the spring. Is this a coincidence? I don't think so. These rams are just super specimens in their environment. They always look 100% thrifty, have no manure on their tails, and have good body condition no matter what time of the year it is. These are the kinds of rams that you are looking for to build your flock.

Control the exposure of your ram to your ewes. You must do this if you want to lamb in the proper time of your growing seasons. You will always hear people say the reason they lamb in winter is because that is when the ram got in with the ewes.

Never run your ewes up against the ram pasture. Rams can be very efficient in gaining access to ewes in heat. They will find a way to those ewes. A neighbor had a ram that swam a pond to get with the ewes. He had January lambs and lost about half of them. Trust me, you do not want to lamb in the winter unless you like pain and agony.

If the ram can not see the ewe or smell them this will help keep them apart. The method we use to control our group of rams from having exposure to our ewes is fairly simple. We gather up our rams and move them to another farm where we have goats and cattle. The rams are far enough away that they can not smell, hear or see the ewe flock.

St. Croix rams are very prolific breeders. One ram can easily breed 60 ewes. The testicle size on the ram is about the same size as a bull, but on a much smaller body.

Once you start building the flock, don't make excuses for your sheep when they do not perform. If a ewe loses her lamb, sell her. You can not afford to give her another chance.

If she develops foot problems, sell her.

If she gets wormy, sell her. Watch the tail area. If it gets runny manure on it, she has got worms. You can look around the flock and see how they stick out like a sore thumb.

Any bad dispositions in rams should not be tolerated. Sell them immediately. We have not had any rams with bad dispositions yet.

All of these problem sheep will pass their inferior genes to their offspring. You can not build a sustainable sheep operation by breeding inferior animals.

One huge advantage of culling a sheep flock versus culling a cattle herd is the cost of doing so. When you cull a cow, you may lose up to $800 per cow, depending on her body condition. With sheep you have so little invested compared to a cow, that it makes culling rather painless to the pocket book.

If you don't cull now you will pay for it later. The inferior genetics will always come back and bite you. You have also wasted the forage that the inferior animal consumed simply because you did not do the right thing.

Once you start ridding your flock of the parasite prone sheep, the parasite load on your pastures will diminish drastically. This is because the wormy sheep are not there to constantly re-infect your pastures with manure infested with worm eggs.

Most sheep have some parasites. The key is to keep sheep that can perform with some parasite load. In nature the wild animals that could not perform with some parasite load in them died. Mother Nature was not going to worm them. Only the resistant ones survived. There are some darn nice prolific deer running our woods that don't get wormed. Think about it, if the parasite killed everything that it took refuge in, it would lose its food source and die.

By being in control of what gets culled for parasites, you are building a future in a profitable sheep flock.

By worming everything, you are weakening your flock. You're not building any future to look forward to in your sheep enterprise. I can not state it any simpler than that.

Chapter 17
The Economics of Pastured Hair Sheep

The Texas Barbados ewes cost us $60 each. We lost 50 of the 200 the first year due to parasites and relocation.

When you move a species that far from an arid environment to a wet environment they struggle big time.

That still left us with $80 invested in each ewe.

Remember, there are sheep for buying, and there are sheep for selling. If you want to build a commodity flock, I personally feel like you can not give over $130 for a ewe. Preferably $60 for ewe lambs is your best choice. I called around a lot and everybody wanted to sell me ewes for $200-500 each. If I had given $300 for ewes and lost 50, that's a $15,000 loss versus the $3000 loss that I took.

This example drives home the importance of buying them right. The remaining ones seem to have adjusted pretty well to their environment and look 100% better than when we received them. They were used to eating sagebrush, and when we turned them into clover/grass pastures, their rumens were not adapted to this new forage.

One of the nice things about sheep is that when you do lose one, it only costs you $60-80, unlike a cow which may cost you $1200. That cost will make your pocket book groan.

We are averaging around $70- 100 per lamb at the local auction. You can pay off your ewe in one year and if she has twins, be in the black the first year. Show me a cow that you can do that with.

You can run seven ewes for what one cow eats. If you

can average 1.5 lambs per ewe for the whole flock, which I think is very conservative, that's 1.5 x 7 ewes (or one cow unit) = 10.5 lambs. If you are selling them for commodity price that is $1050 per cow unit. I have never seen a commodity calf sell for $1050 off the cow.

If you direct market the lambs, the returns look even better. Grassfed lamb is hot right now. People are getting as much as $300 for packaged lamb. That would jump you up to $3150 per cow unit (7 ewes). Folks you can make a good living at this price.

All the figures above are based on lambing once per year. If you go with accelerated lambing, spring and early fall, then your returns could be much greater.

We do not breed for two lambing periods per year. We feel like it is too hard on the ewe. You don't see deer having fawns twice per year. We try and mimic nature as much as we can. It keeps us out of trouble.

I've heard the saying that people raise sheep so that they can afford their cattle habit. I kind of see what they're talking about.

Some of the benefits of selling grassfed lamb are that you have a lot smaller carcass for people to purchase. The St. Croix have smaller carcasses than Dorper hair sheep. They are about even with Katahdin, maybe slightly smaller on the average.

One pleasant surprise is how good the meat is. We eat a lot of lamb because we have it available, and we really enjoy it. Every lamb seems to be tender, whereas with grassfed beef you struggle with tenderness from one carcass to the next. The meat is juicy and does not taste like mutton at all. It's very mild.

Our meat customers love our lamb. Their only complaint is that they eat it up too fast. I just tell them to buy another one!

There is no other meat that lends itself as well to outdoor grilling as grass-finished lamb. It has been delicious every time we have grilled it. If you have a large group of heavy eaters, serve beef. We reserve lamb for special occasions when we have small groups over for dinner.

The trick to grilling grass-finished lamb is to cook it slowly.

Our procedure is to fire up the charcoal and let them burn until the coals are white. The lamb is sprinkled with fresh ground black pepper and sea salt before being placed on the grill. All the white coals are piled into one pile on one side of the grill. The lamb is placed directly over the coals for one minute on each side to sear the exterior surface of the meat.

After the second side has been seared, slide the lamb over to the other side of the grill without turning it. Place the lid on the grill and shut down the air vents until there is only a small crack to keep the coals going.

The size of the meat cut and how well done you like your meat cooked will dictate how long to leave the meat on the grill.

We like ours just a little pink in the center. This ensures that all the natural juices are locked in the meat. An over cooked dry piece of meat is not my idea of a pleasant eating experience. Usually within 5-8 minutes the lamb is ready to eat.

A tip when taking your lamb off the grill, make sure that you place the meat in a warm pan and cover it. Lamb will cool off very quickly when removed from the grill. Let the meat set for five minutes covered. They call it, "Letting it rest." This allows the meat to absorb the juices from the bottom of the pan. I have got to stop talking about grilling lamb. I'm getting hungry!

After four years of the St. Croix, I can honestly say that it is hard to find things to spend money on them. If you are one of those people who like to spend money, these are not the sheep for you. All they require is salt, predator protection, water and forage. We do provide mineral in the non-growing season due to the lack of green forages.

The dog food for our guard dog is probably our biggest expense. I will cover guard dogs later in the book.

As I said before, another benefit with hair sheep is the absence of shearing costs. They shed off each year just like the cattle. In our area it costs more to have them sheared than the wool is worth. It also takes 15-20% of what a sheep consumes to grow all that worthless wool. That is forage wasted that could be growing meat that you can sell.

People report the meat from hair sheep is very mild compared to wool lamb. It is less mutton flavored because it is absent the lanolin from the wool.

If you ask the American consumer what kind of lamb they prefer, they will answer New Zealand lamb most of the time. The reason for this is these are lambs that are 100% grassfed, no grain. Here in America we put them in feedlots like cattle and finish them on grain. Now you have a carcass with excess fat all over it, which turns the consumer off. Grain fat in lamb meat destroys the shelf life. It actually makes the meat rot. Grassfed lamb has tremendous shelf life compared to grain-fed lamb.

The grassfed lamb is high in omega 3 and CLA (Conjugated Linoleic Acid), which is a cancer fighting agent. The American lamb board is screaming about all the imported New Zealand and Australian lamb coming into this country. Heck, if the USA sheep producers would start finishing their lamb on grass, which is what the customer is demanding, we could compete with the imported lamb and get our market share also.

Chapter18
Adding Brush Goats

I must admit we were very skeptical about getting into goats. I had read a lot of horror stories about how tough they were to keep in. We have all heard the saying, "There never was a fence built that a goat could not go through."

I had dreams of them getting out on the highway, eating the neighbor's garden, leaving and not coming back.

Goats can be very vulnerable to parasites. I did not want an animal that was going to require constant worming to keep alive.

The farm where we put them had an old barn, so their shelter needs were met. We figured we could work around buying any additional handling facilities.

We dove in anyway and are learning as we go that a lot of our fears are just not there.

We do not provide any fancy shelter and have built no extra handling facilities other than backing up our corral panels with regular woven wire to keep the goats from crawling under them when we're sorting or working them. The old barn that is on the leased farm is what they use when we have cold rain. A goat needs to have a place to dry off or you will have some sick goats. Sheep do not require any shelter.

We have not had to worm our goats yet, because we let them wander over a large area, and they are not confined. By confining them they get exposed to the parasite buildup in those areas.

We typically keep around 20 goats. If we had hundreds,

they would have to be confined to paddocks and rotated to avoid parasite infestation.

Initially we thought we would rotationally graze them like the cattle. After some first hand experience, we prefer to let them wander. They seek out all the blackberry patches, thorn trees, and woody plants that they can reach. They have no ill effect on the forages, and actually complement them by removing the woody plants. In a way they are rotating, but they do it on their own.

We bought some 30 to 50 dollar does and bred them to a Boer billy for a better meat carcass.

The goats are very prolific. They seem to average twins after their second year. Some years we get two crops of kids, one in July, one in November. We do not take out the billy from the does.

They are fed no grain, no hay, just green forage in the growing season and stockpiled forage in the winter.

If you have a cedar problem you can minimize it with goats. This past winter the goats ate mostly shrub bark, and picked through the stockpiled grass, but when March arrived they literally destroyed the cedar trees. They stripped all the bark off of the trunk as high as they could reach. They even climbed up in the bigger trees and stripped the limbs. I can not explain why at this particular time of the year they attacked the cedars, but they sure did. Cedar bark has a lot of tannin in it. I wonder if cedar tannin is nature's way of eliminating parasites in them. All I know is I like goats that eat cedars.

There is nothing more entertaining to me than to watch a flock of goats surround a big multi-flora rose bush and start stripping the leaves off of it. They look like a wagon wheel formed out of goats just chomping down. When they leave, the plant looks like a skeleton.

I prefer this method over chemical sprays. The goats will keep coming back and eating the fresh leaves as they try to emerge. They also eat the actual limbs if nothing else is available.

I've grown to love goats for all of the above reasons. We have changed our attitude about eliminating all this brush; you want enough goats to just control it. If you stock too heavy they will kill all the brush, which eliminates their food source.

I know it sounds crazy but I like a little brush in our pastures now. Remember, diversity is king!

We castrate our goats at weaning time so that we can concentrate on the Boer meat goat cross. There are people who would prefer to not have them castrated, and there is a market for intact billies also. We are keeping the best to grow our flock. Our goal is to have a flock for each farm to keep the brush under control. We do not kid in the winter, only in the spring, summer and early fall.

If you have a nice vehicle, do not park it where the goats have access to it. When you come back to your vehicle there will probably be several goats standing on the hood or the top of the vehicle.

We have an old barn that has a hay mow in it. The ladder that gives you access to the loft is just a remnant of a ladder with most of the steps missing. The goats are always figuring a way to climb it. They will be standing in the hay mow door looking down at you when you drive up. I had one young goat that stayed up there for two days before it figured out how to get down.

We do have very effective perimeter fence to keep them in - a minimum of five hi-tensile wires. One farm has seven hi-tensile wires on the perimeter, and every other wire is hot.

The bottom four wires need to be closely spaced to prevent the goats from slipping through them. If a goat can get his head through, the rest of him will follow. We have had good luck with 4-6" spacings. The whole key is to have a hot, hot fence. Once you knock the stuffing out of a goat, they are a whole lot more skittish the next time about challenging the fence.

We have tried to hold them with three hi-tensile wires in a paddock that was holding sheep with five. Those goats would walk right through the fence without missing a step. This fence had 4500 volts on it, and it never bothered them a bit.

We have one farm where we are pushing around 8000 volts with ground wires in between. The wires are on 6" spacing and very tight.

I still remember the look on this particular rogue bunch of goats when I turned them into this farm. They had previously been wading through three strands of 4500 volts with no ground wires mixed in. The first goat walked up to the fence and stuck his head through the wire. His momentum was immediately backwards. It sounded like you had shot a 22 magnum rifle.

Ker-pow.

That goat fell backwards and shook himself off in disbelief and never approached the fence after that.

If any wire has a little sag to it, that is an open invitation for a goat. I actually saw one try to jump through it. Even though all four feet were off the ground it still made this huge pop, because he got grounded by hitting the ground wire. The goat let out a scream and was quickly broken from trying that again. I can not stress enough the importance of a good ground system.

This particular farm has 12-6 foot galvanized ground rods by the charger and an additional ground rod every 1/4 mile around the perimeter of the farm. These additional ground rods are driven right under the perimeter fence and have an insulated hot wire connected to the ground wire on the fence.

We have a good ranching friend in Oklahoma who has a large flock of goats. We happened to be going by his ranch on our way to Dallas one summer. He gave us a farm tour of his 10,000 acre ranch and then casually asked us if we would like to help them with a goat round-up. That sounded like fun to us.

He needed to worm the kids and castrate the males. He got together six other ranch hands, and we were set for the goat roundup. Little did we know what we were in for. I'm thinking a nice leisurely thirty minute walk down across the pasture, then push the flock through a gate.

He had 3000 goats of which half were six-week-old kids.

I thought we had thick brush in Missouri. Our brush can not hold a candle to Oklahoma brush. Everything has a thorn on it!

The goats were running on 80 acres of mainly saw briar brush. There is a reason it is called saw briar. It grabs you and saws on your meat.

The goats were spread over the entire 80 acres of brush like you sprinkled them out of a salt shaker. We all fanned out on the farthest side away from the gate where we needed to take them. The older goats would move forward pretty well with pressure, but most of the kids would just stand and scream. It sounded like 10,000 three-year-old kids crying their eyes out.

I will never forget the first saw briar bush I tried to stomp through while heading off a group of goats. This saw briar got me from all sides. I was pinned. Trust me, you don't want to move when one of those buggers grabs you.

I had three-quarter-inch thorn needles sticking me from all sides. I could not move. I very carefully pulled each one of the thorn needles out of my flesh and finally wiggled down on the ground and crawled out of that bush. I have to be truthful though, I was on an ATV most of the time. My wife was on foot the whole roundup.

Before you all start poking fun at me for not letting my wife have the ATV, let me explain. This ATV was not your typical easy-to-drive ATV, plus the terrain was very steep, more like an obstacle course. Jan did not feel comfortable with the idea of driving it over this brushy steep terrain.

The 80 acres had a small creek one foot wide that ran right through the middle of it. If you were lucky enough to get 100 or so goat kids to the creek they would not cross it.

The procedure to get them to cross was kind of cute. You held pressure on the kids until one brave soul would jump the creek, then the other 99 would jump the creek, one at a time. Most of the kids would circle back through the brush where they came from at the speed of rabbits.

Four hours later we had most of the goats into the other pasture. It was quite a sight. There must have been 100 or more kids still screaming back in the saw briar brush pasture. Even though we had on jeans, we were all pretty cut up.

My wife got the worst of it, and she was completely exhausted. It was an experience we will never forget.

Chapter 19
Adding Tamworth Grazing Pigs

We added Tamworth grazing pigs to our farm several years ago after reading about how well they performed on pasture.

We started out with two thirty-pound gilts and one small boar.

The first thing we did was break them to hot wire. We had them in a board-sided pen that was grown up in weeds. It was amazing to watch the little pigs tear into the weeds.

After they riddled the weeds, I strung a piece of polywire and positioned it around the pen about six inches off the ground. Before I could get the power turned on to the polywire the young boar was chewing on it like a sandwich. Finally, I threw in a couple of old apples, and they started hogging them down. This gave me time to turn on the power.

I crept back to the pen and sat down to watch the show.

The apples were gone in no time, and the young boar decided to have another polywire sandwich. The boar opened his mouth and completely took the wire in clear to the back of his mouth like a horse bridle. I can still see the look of horror in the boar's eyes when he chomped down on the electrified polywire.

The boar let out a squeal that you could have heard for a mile. It looked like you shot him out of a slingshot. He actually shot back so fast that he bounced off the opposite wall of the pen and ran in his pig house barrel still squealing. The two young gilts ran in the barrel and joined him.

I came back about an hour later and they were all still in the barrel cautiously looking outside. This one experience broke the boar, and the gilts were schooled the next day.

Pigs are the easiest thing in the world to break to hot wire. They do not have any thick hair to insulate them from the shock. They also investigate everything with their moist noses, which are pretty conductive. We never had the pigs try a hot wire fence after that. Our fencer puts out about 7-8000 volts. I ache for awhile when I touch it. I can not imagine clamping down on it with my teeth in my moist mouth.

I had a friend who raised hogs in pens. He had one boar that he kept separated from the sows with several hot wires. When a gilt came in heat, he would stand back about 20 feet from the hot wire and start squealing. Then the boar would take off running right at the hot wire and shoot through it. He knew that it was going to hurt when he hit it, but that sow was worth it!

I don't think there has been an animal that we have introduced onto our farm that we have enjoyed any more than the pigs. My wife thought I had totally lost my mind when I asked her about getting some pigs.

All she could envision was mud, and stinky smelly animals rooting up everything. Well, it has not turned out that way at all.

The Tamworths are truly the best grazing pig I have ever been around. They can thrive on pasture alone. They do not smell at all when left on pasture. We feed our pigs garden scraps to keep them tame. In the winter time when everything is covered with snow or ice, they get a couple pounds of shelled corn each. The first time we heard them eating walnuts, my wife commented, "Sounds like they are eating rocks."

We have never wormed or given any of our pigs shots. As long as you give them free access to clean pasture, they stay perfectly healthy.

We did ring the noses of our first set of pigs because we had read so many horror stories about how they would tear up your pastures by rooting. We never ringed any more pigs after that first year. As long as the pigs are not cooped up in a small area they do not root. At least our Tamworth don't. They are true grazers and are very content to browse the pastures picking out their meal.

If you decide to ring their noses, use a wire snare to hold them. We make one out of 12 gauge wire to run into the pig's mouth and pull the loop tight against the top part of its mouth with a pair of pliers. This sure makes it easier to hold them while you are putting several rings in their noses.

Our ponds have one hot wire that is 32" high to keep out the cattle. The pigs can easily walk under this wire and get a drink or cool off. We have not seen any pond damage by the pigs rooting in them.

They will eat mice, snakes, grasshoppers, crickets or anything that smells good to them. Their nose is amazing at how well they can scent things from a distance.

We have a guard dog that patrols the farm that has the hogs, hairsheep, goats, chickens, and cattle on it. I can not use a self feeder for the guard dog on this farm because the pigs scent out the dog feed. We have to hand feed this guard dog daily to keep the pigs from eating the dog food. The guard dog is no match for a Tamworth that smells dog feed.

When we first introduced the Tamworths to the cattle herd, it was a circus. The whole herd of cattle bunched up together in one corner of the pasture, staring and mooing at the hogs. They had never seen a hog in their life, and were not sure if they were predators or what.

The herd bull was right out in the front of the cow herd with his head down waiting on the hogs to move closer. The boar hog's attitude was like, "What's up folks?" as he just sauntered up to meet the cattle herd.

The herd bull immediately charged the boar. The boar just nimbly side stepped the bull and kept going forward into the cow herd!

The boar had made it through the first line of the mob's

defense and was still just trying to be friendly with the cows. The cows started butting him around and he would let out a squeal. It took about a day and the cows did not pay any attention to the hogs after that. The hogs would actually lay down right in the middle of the resting cow herd.

The Tamworths kind of nose through the cow pies spreading them out, and they love rolling in them in hot weather. I guess a fresh runny cow pie is kind of like a mud hole to a hog. I have seen them completely covered from head to tail with green manure.

My wife, Jan, noticed that the Tamworths would cruise around the resting cow herd and walk up to a cow that was lying down to nudge them a little. Guess what a cow does when she stands up? You guessed it, a fresh runny virgin cow pie to roll in. It is hard to believe that a Tamworth could figure out how they can get their next runny cow pie to roll in.

The way we bonded the hogs to the cattle the first year was to feed them their little ration of corn every other day with the cattle. The hogs learned to stay with the cattle because that is where they were fed.

When we introduced the horses into our mob of cattle and hogs, the horses freaked out. The horses were bound and determined to run the hogs into the ground.

I will never forget that day. It was about 95 degrees F and we made a paddock move that covered about a quarter mile down a lane. The horses had already been turned into the fresh paddock when the cattle mob and hogs got there. The horses never looked up when the cattle mob entered the fresh paddock. About that time one of the hogs just walked up and grunted at one of the horses. The horse farted real loud, let out a shrill snort, and took off running like its tail was on fire. The other horses took off running and blowing their noses with their heads arched back.

The horses immediately surrounded the hogs and started closing the circle around them.

The hogs were like, "What's up?"

The horses would make a mad dash at the hogs, biting and kicking at them. Finally the horses put enough pressure on the hogs that they started running. Well, a hog can not outrun a horse. Finally, after the hogs had been run for 15-20 minutes the hogs all came together and laid down in a tight circle facing the horses. They were just too hot to run anymore and figured safety was in numbers.

The horses circled them snorting and biting at them for 30 minutes. The hogs finally laid completely down on the ground and went to sleep. The horses went back to grazing and we never had any more major problems with them bothering the hogs.

I think animals just have to work it out socially, and things are fine after that. If I had intervened and separated the hogs and horses they would have never worked it out for themselves.

This kind of gives you an idea of how tough these Tamworths are. Can you imagine running a confinement hog in 95 degree F heat for 20-30 minutes? You would have a pasture full of dead hogs. Most of them would die from the stress and heat.

When we first put the hogs with the cattle mob we got some stares from people passing by in their cars. The farm that they graze has a blacktop road frontage that is heavily traveled. There has been more than one time that I saw people go by and their brake lights came on when they saw a pig grazing along with the cattle mob.

The local coffee shop was having a feast day on the subject according to my neighbor. "Did you see what that crazy Judy is up to now? He's got hogs mixed in with his cattle!"

One fellow replied, "How in the heck does he keep them hogs with the cattle?"

My neighbor could not stand being quiet any more. He replied, "I know how he keeps those hogs with the cattle." The coffee shop fell silent waiting for his answer. My neighbor replied, "Them hogs think they are cows!"

I wish I could have been a fly on the wall of that coffee shop. What a wonderful answer.

It is a true treat to watch them graze in the pasture. They walk along at a steady pace with their heads bobbing just enough to take in the top 2-3"of clover or whatever the target may be. They graze at lightning speed, just taking the best tender part of the plant as their snout passes over it.

They are also amazing to watch eating acorns at a steady walk. They are chewing up an acorn in their mouth while at the same time their nose is moving leaves, sniffing out the next victim. The shells of the acorns are sifted out the bottom of their mouths while the inner seed is chewed up and swallowed. Think about that, chewing, sifting hulls, seeking the next acorn at a steady walk, all with lightning precision.

After one winter of keeping hogs, we decided to buy feeder pigs in the early spring. This way we do not have to feed the breeding stock through the winter. Corn gets expensive at five dollars a bushel.

From now on, our spring feeder pigs will be butchered in the late fall after they have harvested the acorn, hickory nut, walnut and weed seed crop.

I will miss pigging time though. Our Tamworth guilts would go out in the brush at pigging time and be gone for about a week. We would anxiously wait for the sow to bring them back to the barn area so that we could count how many she had.

They would come walking in after about a week with six to eight little pink piglets. Nothing is cuter than a newborn baby pig. They look like baby mice.

One word of caution when Tamworth guilts pig, stay away from them until the baby pigs are several weeks old. This is the only time guilts are aggressive toward people. The first guilt that pigged was gone for eight days.

One of the questions we get is, "How do you keep the coyotes from killing the baby pigs out in the pasture?"

My answer is, "I pity the poor coyote that tries to get a baby pig away from a mature Tamworth sow!"

131

My neighbor and I had gone to the farm together. I went to move the mob. He went to the barn to gather our eggs. When I got back to the truck, he was sitting in the truck with the windows rolled up. It was July and 95 degrees F! His eyes were bugged out and he would not roll down the window. I could not imagine what was wrong with him.

I asked him, "What you doing in the truck with the windows rolled up?"

He immediately answered, "I went to gather those eggs in the barn and that old sow was laying in there suckling those baby pigs. I was on top of her before I could stop my momentum. I did not know I could run that fast."

My old neighbor weighs about 270 lbs, but he was a flash that day. After that my neighbor always peeked around the corner before going in the barn. That same day though, the sow walked right up to me with the baby pigs in tow and never showed any aggression toward me at all.

I had a huge field of cockleburs that were the result of the previous landowner's cropping practices. He planted soybeans every year in this field and combined the mature cockleburs right out on the ground. I remember driving past the field and not being able to tell whether it was a bean field or a cocklebur field.

We put the mob of cattle on it and strip grazed it off very tightly. The cocklebur plants were about two to four feet tall and the actual burs were not mature yet. When you squeezed a bur together with your fingers they were not hardened enough to stick you. The cattle ate every plant to the ground, but in the process knocked off thousands of immature burs on the ground.

I went out to move the mob one morning to their next strip, and the pigs were having a party on the loose burs spread on the ground. They were shoveling them up with their mouths like a vacuum cleaner. I was amazed that the hogs would eat them.

I cut a bur in half with my pocket knife to see what the

center looked like. There were two juicy little seeds in the middle of it.

They were also eating the juicy cocklebur stalks that the cows had pulled out of the ground while trying to eat them. Here were two different animal species just enjoying the heck out of themselves ridding the pasture of burs! No chemical spray, no fossil-fuel-consuming tractor and mower, no labor on our part. The animals were performing a service and gaining weight in the process.

Hogs are the easiest thing in the world to load or catch. Just throw a little corn in the trailer and close the door after they charge in. Don't go out and buy an expensive hog house like I did. What a waste of money.

Hogs do fine until the weather gets really cold, then they need a bale of hay or something to make a nest in. They do need access to shade to stay cool when it gets hot. We have an old barn that they get in when we have snow and ice. They are a very low maintenance animal during the growing season.

When you castrate the small pigs, get inside something that will protect you from the sow. We throw some shelled corn down in the cattle alley to lure all the pigs in and block off the exits. Next we take a piece of plywood that fits the width of the cattle alley and sort the big hogs out of the alley. This leaves us with a nice strong steel alley between us and the large hogs. A little pig squealing is a distress call for help. The older hogs go nuts, making woofing noises at you. Once the squealing stops, the older hogs completely ignore you.

We catch the piglets and castrate them while they are five to ten lbs. They are still a handful to hold, just solid wiggling muscle.

Never leave children unattended around hogs. The first thing a kid wants to do is pet one. This can be disaster. A child holding out their fingers to pet one is an invitation to a hog to take off their fingers. The hog thinks that the kid is offering it something to eat (fingers).

Never ever hand feed a hog for any reason. Their jaws

are as strong as a Pit Bull's. I will not even allow our hogs to sniff around our clothing. You never know what kind of food you may have spilled on your pants. The Tamworths are wonderful docile hogs, just respect them for what they are. We sure enjoy them.

With the first Tamworth we processed, the meat was a genuine surprise. When we opened the first package of sausage, it looked like hamburger. It is the best sausage that I ever ate. There was barely enough fat to keep the sausage from sticking when we cooked it. There was no grease to dispose of after cooking. The meat does not look at all like the pale washed out tasteless pork that you buy in the store. It has a very rich, reddish pink color to the meat, and the taste is wonderful.

We had some bacon cured from one of the hogs. It's wonderful stuff.

Chapter 20
Selecting the Livestock Guardian Dog

When selecting a livestock guardian dog you need to be cautious. Inquire about the background of the pup.

Are the parents working guard dogs? Ask the breeder for references. Look around the breeder's facilities. Does he have the dogs with the sheep? This is a must in my opinion.

Good working guard dogs will pass on their guarding instincts to the pups most of the time. There are so many people raising these dogs for pets that a lot of them have never even been around any livestock.

When you walk up to a group of puppies to inspect them, don't pick one that is timid or cowers away from you. If the puppy walks up to you wiggling it's tail, it has a lot of self confidence and will be a dog that you can handle.

One of our female dogs, Dusty, is very timid. Dusty will not let you touch her or catch her. She is now four years old, and it is a real pain if you have to catch her for some reason. The only way we can catch her is to lead the sheep flock into the corral with her following them. Dusty is a wonderful guard dog and works very hard for us, she just never gets a pat on the head for a job well done. I do give her a lot of "good girl Dusty" praises.

Don't over spend for your pup. You should be able to find a good pup for $200. We paid $250 for both of our last two pups. Our big male Great Pyrenees was given to us by folks who had goats and had to sell out. He has turned out to be a wonderful dog after getting acclimated to our sheep flock.

When we first got him we penned him up in an electro-net pen in the sheep pasture. The pen was located right next to the sheep mineral feeder. This gave the dog lots of exposure to his future family.

Things were fine the first day we turned him out with the sheep and the other female guard dog. The second day I found a dead lamb that was half eaten. I was horrified and couldn't believe that the new guard dog had killed a lamb and eaten part of it. I told myself that I had not seen the dog actually kill the lamb, so therefore he was innocent until proven guilty. The lamb could have died of some natural cause.

You could tell from what was left of the lamb that it was a nice healthy looking lamb when it died. The previous owners swore that the dog had never shown any aggression toward their goats. That was two years ago and we have never found another dead lamb eaten since.

Check the internet for different guardian dog web sites. A lot of good ones tell the history and what you can expect from their breed. These dogs get big and can become a nuisance if they don't have the guarding instinct.

I have seen a lot of guard dogs that would never be with the flock, but instead were usually at the neighbor's. These dogs are worthless for guardian animals. They are friendly to people and are very aggressive to predators and anything perceived as a predator.

We were very fortunate with our first guard dog pup. We have a good friend who lives in Indiana who bought two pups from a breeder who had working guard dog parents. He raised them together with his sheep until they were about six to seven months old, then he brought us one of them.

Some breeds of guard dogs tend to be more aggressive than others. We only have experience with the Great Pyrenees. The Great Pyrenees puts on a good bluff by barking loudly. Personally we don't want any dog that we have to worry about biting someone. There's too much risk of a lawsuit, plus we have to be able to handle the dog ourselves.

Guardian dogs don't have to be mean to be effective predator deterrents. All they have to do is be with the flock and be heard to be effective most of the time. The only other animal in the wild kingdom with a louder decibel reading than the Great Pyrenees is the African lion.

We have one farm that the large flock of sheep are grazed on. We have two guard dogs that live with that flock. On another farm we graze goats, and also put our rams on this farm to keep them from breeding our ewes. Placing the rams on a different farm allows us to lamb in May. We have one lone guard dog that lives with the goats and rams.

The smaller the paddock that your flock is on each day, the easier it is for your guard dog to protect them. The two farms are far enough apart that the guard dogs can not hear each other or mix with each other.

We do not have guard dogs protecting the cattle herd exclusively. On the goat and ram sheep farm, the guard dog does make her rounds through the cattle herd sometimes, but mainly stays with the goats and rams.

With the High Density Grazing method that is used with the cattle, guard dogs are not needed. By having your cattle bunched up, the predators will be kept out by the large mob. If we could figure out how to keep the sheep with the cattle mob on a daily move basis, the guard dogs could possibly be eliminated.

A lot of people who visit our farms are surprised that we do not have a Border Collie for gathering or moving our flock. Everybody told us that we would have to have a Border Collie to handle the sheep. We went ahead and bought the sheep anyway without one. I had my doubts when we first got a sizable flock that we would be able to control them without a Border Collie.

My argument for not having a Border Collie was that it was another mouth to feed, and they don't give those dogs away. You have to really know what you are doing to train a Border Collie. A well trained Border Collie can cost $1000 or more. It would take a lot of lambs to pay for such a dog. I could also envision our $1000 Border Collie ending up as guard dog fodder.

The new sheep flock was pretty skeptical of us when we first purchased them. The hair sheep flock started to let their guard down some when we gave them a treat of shelled corn. We only gave them a cup full. Within no time we had the flock to the point that they would follow us to St. Louis if we shook a few kernels of corn in a bucket.

We now have the sheep trained to move by just saying, "sheep, sheep, sheep," real fast. I must admit there have been a couple of times when a Border Collie would have come in handy to sort out a particular sheep or group of sheep. We have learned to get by fine without one. Just don't ask me to raise sheep without guard dogs!

Chapter 21
Training and Care of the Livestock Guardian Dog

People are always asking us how we trained our guard dogs to guard the sheep.

When we first got our puppies, we never separated them from the sheep. They were fed with the sheep and slept with the sheep.

The first guard dog we purchased was a Great Pyrenees. Her name was given to her by our friends' little four-year-old boy. I remember them bringing her to us, and my wife had a name tag made up to put on her collar along with our name and phone number. That way if she wandered off at least people would know who to call.

My wife had misunderstood Roo's name and put the name Boo on the collar nameplate. The little boy was shocked when he arrived and saw the nameplate had Boo in place of Roo! He did not let my wife live that down for quite ahile.

At the time we got Roo we had just started our small flock of St. Croix. They had started lambing in a paddock made out of electro-net. The sheep had never been around any guard dogs and viewed her as a predator. So, I put up an enclosure in the pasture and put the ewes and lambs in it. Next I put Roo in the enclosure in her separate pen, and I put the sheep's water and mineral right next to Roo's pen.

After about one week I turned Roo out with the sheep. The sheep stood in one corner and stomped their feet, warning her not to come any closer. Roo had been raised around sheep and just sauntered over next to them. Two ewes charged her at the same time and knocked her down on her side. When she tried to get up, they knocked her down again. Finally she just stayed down on her stomach in a very submissive posture. Never once did she offer to retaliate toward the sheep.

This was my first good sign that I had a good guard dog. They must never retaliate unless a sheep tries to take their food. A guard dog must guard its food, but should not physically harm a sheep while doing so. A quick jaw snapping and growling episode makes the sheep back off real quick from the dog food bowl.

After several days the ewes did not pay any attention to Roo unless she got too close to their lambs. The lambs would actually go over and climb on Roo. She would just lay there and maybe lick on them or sniff them.

As night approached, any strange noises would cause Roo to bark with very forceful warnings. Any house dog in the distance that barked instantly drew Roo's attention.

Next, I introduced Roo to the cattle herd. I turned the sheep in with 90 cow/calf pairs. Roo was very protective of her sheep. Any time the cattle got too close to the sheep, she would position herself between the sheep and them. The cattle were alarmed with the big white dog at first, but after a couple days never paid any attention to her.

I remember the first time I fed Roo in the cow pasture. The whole herd came up to see what goodies I had. One old boss cow just walked up to Roo and was going to run her off her dog feed. Boy did that cow get a surprise. Roo instantly gave that cow a face full of teeth snapping and growling. The cow went backwards as fast as it could and kind of stood there in awe. Roo never touched the cow's head, but persuaded her that her head did not belong in her dog food bowl.

We trained two pups using the "isolated with ewe" technique. These pups have the same parents as Roo. We penned them up by themselves and put an old ewe in each pen. The reason for penning them separately is that the pups will have no choice but to bond onto the ewe. There is not another puppy to buddy up with. The pen was made of electro-net fencing. Basically it is woven electrified netting that keeps livestock in. The pups learned real quick that hot wire was to be respected.

After a month of being penned with their ewe, they bonded. The pups were turned loose with Roo and the flock. The pups followed Roo around and she basically reprimanded them if they did something not to her liking. I never saw them run a sheep or stalk a sheep.

If you have a guard dog that stalks a sheep, this is an inferior instinct that you can not cure. Give the dog away to somebody who needs a family pet.

This method of training pups is a pain because you have a separate set of animals that need feed and water. The next pup we get will go right in with the older dogs under a watchful eye. If this method works, it sure would save a lot of labor bonding the pup to sheep.

The guard dog must be submissive around the sheep and also to you. Never allow the dog to be the boss over you. This is dangerous for you and anybody else who would happen to come in the pasture. If your guard dog comes up to you and is rough, knocking up against you, basically not respecting you, then you need to give him a lesson.

We reach down and secure a hold on the dog. We gently push him down on the ground on his back. Hold the dog in that position for awhile.

If he starts squirming, let him up. If you move slowly and do not get excited, most dogs will just lay there. You are showing them that you are the boss and the dog has to be submissive to you. Any dog that lies on his back is in a submissive posture. If the dog continues showing no respect, continue the treatment.

Don't love on or pet your dog constantly or all you will have is a pet. The dog will bond onto you instead of the sheep. When you leave the sheep pasture and go to the house you will have a companion, not a guard dog.

You want that dog to think that it is a sheep. This is easier said than done because you have to guard your emotions especially when they are puppies.

When these pups are small, they look like little white fur balls with little black chunks of coal for eyeballs. There is hardly anything cuter than a Great Pyrenees puppy. They bounce, jump, wag their tails, have little pink tongues hanging out their mouths, and would just love to be petted by you. It takes a strong soul to resist this temptation.

I can not allow my wife to go around the pups. She admits she is hopelessly weak in this regard. She can not resist the temptation to love on them. I know that I have to resist petting them excessively as well.

Look at it this way, you are saving their life by not petting them. These dogs are bred to protect. If you pet them they will end up as a pet or something worse, instead of what they were bred to do.

Depending on who you talk to, there are different schools of thought on how much handling the pups need. My belief is that you have to catch them from time to time to examine them or to move them to another farm.

We don't teach them to walk on a leash.

I had a lady vet chew me out when we brought our pups in to get them fixed. She started out by saying, "Sir you should always teach a dog to walk on a leash. You are doing a huge disservice to your dog and anybody who wants to walk your dog by not teaching them this basic process."

She may have been right, but we do not own these dogs so that we can walk them around on a leash. We teach them their names and to come when called. That's it.

We worm our dogs by mixing the wormer with a chunk of food, usually a piece of wadded up bread covered with a little meat grease or butter with the wormer capsule in the middle of the wad. It is amazing how healthy the dogs stay on pasture. They are being constantly moved each week to a fresh rested paddock. We believe that constant movement to fresh rested ground helps keep the dogs from being infested by parasites.

We do not shear our guard dogs. The long hair protects their skin from the UV rays of the sun. Their skin is very pink under the long hair and can not take direct sunlight. If they get a mat of old hair hanging from their body, we will cut it off.

Never beat a dog for anything they do wrong. There are no excuses for beating. This will only alienate the dog from you and make it harder for you to teach it right from wrong. These dogs are smart. All you have to do is lower your voice and loudly say, "NO." Dogs are so perceptive of our body language and voice tone. They get the message real quick that you are not happy with their actions. It does no good to holler at them unless you catch them in the act of the bad deed.

Treat them like your working partner, because that is just what they are. They want to please you and do a good job for you. When I feed the dogs or go to check the sheep, I give them one pat and tell them they are doing a good job. I think a guard dog needs to know that you appreciate the job they are doing for you.

Roo comes up for her pat on the head then she goes back to her sheep after spinning in circles and acting like a puppy after being petted. I think it is her way of saying thanks for the pat on the head.

If you have a guard dog puppy that wants to chase the sheep, they must be taught early that this is not acceptable behavior. Most of the time you can holler at them, "NO!" and they will stop. If they continue to do this, fasten a log chain to their collar and tie a small car tire or log to the chain. This slows the pup down enough that they get tired of running sheep with all this weight tied to them.

Guard dogs stake out their area by walking the boundaries and marking it with their scent to keep out predators. One of the things a guard dog does not recognize is property lines. So they may try and stake out a huge area if they are allowed to. This is where a lot of guard dogs get in trouble. They will get hit by a car, shot, end up at neighbor's house, or be stolen or lost. Any one of these results is catastrophic for your guard dog.

Once they get to roaming, it is hard to break them of it. If they are allowed to roam, they are not guarding the sheep. I don't know how many sheep operations I've been to that when asked where their guard dog is the reply is, "He's at the neighbor's house." Well, they're not doing you any good at the neighbor's house. You might as well get rid of them. The last thing you need is a feed bill that is not paying it's way.

Our two pups got into the roaming mode simply because there are three houses that adjoin the farm that they are kept on. Each house has house dogs that bark, and kids in the yard playing. It sounds like fun for a pup.

We added an additional hot wire on the bottom of the fence around the perimeter of the farm, and we also tied a 14" tire and log chain to the pups. This has really worked. There are no more calls from neighbors saying they have a pair of beautiful white pups at their house.

The tire method sounds cruel, but it works. The pups can still make their rounds and protect the sheep, but it really cuts down on their excessive roaming.

There are two methods that you can use for feeding your dog. Either hand feed them each day or give them access to a self feeder. There are pros and cons for each method.

By hand feeding, you get to check the sheep and have contact with your dog. If you have a shy dog this is a good way to let them know that you are nothing to fear. They soon learn you are their food wagon. The downside of hand feeding is that you have to do it every day. I don't mind it because I usually am rotating cattle or something anyway.

I do have a self feeder. It holds 50 lbs of dog food in a weather tight feeder. I built a small 4' x 5' wire cattle panel pen on wooden skids that has the dog feeder mounted inside it. The skids are made out of treated landscaping timbers with a 45 degree angle cut on the front of them, which allows the feeder to be pulled over small obstacles without getting hung on them.

The entrance is an 8" space at the bottom that allows the dogs to slide under to get to the feeder. The top surface of the dog food entrance has a board fastened to it to keep the dogs from rubbing on the cattle panels when crawling under it. This keeps the cattle and sheep out of the dog feeder. I can hook it to our ATV and pull it to the next paddock with ease.

We really had a hard time getting one guard dog to go in the feeder. She was convinced that she would get shocked by sliding under the board I guess. After lots of meat scraps being consumed by her buddy in the self feeder, she finally overcame her fear. This self feeder will hold enough dog food for five to nine days depending on the weather. The colder it is, the more food they consume. In the warm weather season, they slack up on the feed a lot.

We always position the dog self feeder door facing south. This keeps most rain from entering the lip of the feeder, which is where the dog food trickles down for the dog to eat.

Do not let your guard dog run out of dog food. They are working hard for you. They deserve to have access to all the

good dry dog food they can eat.

We sold some bred St. Croix ewe lambs to a fellow several years ago. He was having a wonderful time with his sheep. They were lambing and everything was fine. One day I got a call from him saying that his own dogs had got into his sheep and killed several, wounding the rest. He had to get rid of his own family pet dogs because they did not have a guard dog with the sheep. If there had been a guard dog lying up there among those ewes and their new lambs, there would have been a different outcome.

You may be wondering if guard dogs are worth the fuss or not to train and own. Right now we would not be in the sheep and goat business without one. It allows you to sleep well at night knowing that you have 24 hour protection of your flock. There is nothing more defenseless than a baby bleating lamb. It is a food call for coyotes, bobcats, foxes and packs of dogs. We don't worry about it because of our guard dogs.

I talked with an old seasoned sheep rancher from out west about comparing guard dogs versus llamas or donkeys. My interest in llamas and donkeys was the fact that we could get away from the dog food bill. Llamas and donkeys could graze right with the sheep. He told me that he had the same idea several years back.

He bought some llamas to put in with the sheep flock and got rid of the guard dogs. Llamas and dogs do not mix at all. The rancher went over to check on his sheep one night and found dead sheep lying all over the pasture, many more mutilated sheep dying from their wounds. The llama was hiding up in the barn from the neighboring dog pack that had attacked the flock.

The rancher made the comment that a person could buy a lot of dog food for what it cost to replace all those dead and dying sheep. A couple of lambs would pay for the dog food for a whole year for one dog. His parting comment was, "Son, buy a guard dog and sleep well at night." Pretty good advice from a seasoned sheep rancher.

Our landowner called us one night and gave us the story of Roo in action with her flock that afternoon. We had a neighboring dog get into the sheep pasture and Roo went into action immediately. Roo took off running full speed and barking toward the dog with her hackles sticking straight up. All the sheep immediately ran up to the top of the hill behind her and got in a tight circle. Roo kept herself positioned between the flock and the stray dog, viciously barking as the stray dog circled the sheep.

Roo did not attack the dog; she stayed with the sheep protecting them. If there had been two dogs and Roo went to fight one of them, guess where the other dog would have been? This stray dog probably figured there was an easier dinner somewhere else, rather than dealing with this large white dog.

We have lots of coyotes on our farms. When they start howling at night the guard dogs go nuts. They start running in the coyotes' direction barking viciously. The coyotes shut up. For a coyote to make the kill, they have to go through a stalking sequence, and the barking of the guard dog disrupts their stalking sequence.

I made a serious mistake two years ago with some weaned ram lambs. We had a farm that had some good forage and sheep fence on it that was being used for our mature rams. When we weaned the ram lambs, we added them to the old rams figuring everything would be fine. No guard dogs were available for this farm. Within one night we lost six ram lambs to coyotes.

At $175 each, which was what we were selling pure St. Croix rams for at that time, that was a $1050 hit in one night. It surely could have paid for some dog food with that chunk of money!

All that was left of the lambs was the hide and the hooves. You could not have done a better job of skinning the lambs than those coyotes did. It was amazing. The skin was peeled off without any holes chewed in it. Man those coyotes love their lamb. We put a guard dog in the next day and that was the end of the lamb killing. I was naive enough to think that the older rams would be pretty good protection for the younger ram lambs. I was dead wrong.

One other advantage of having a guard dog is that they are present on your farm at all times. This is a deterrent for vandalism that may take place in your absence. Several of my landowners were very concerned when I asked them about placing a guard dog on their farms to protect the sheep. Their concern was that the guard dog would attack their dog when they came out to the farm. I told them to keep their distance from the sheep when they had their dog with them. This has worked fine. The guard dogs do not bother the owners' dogs. They even come up and sniff around them. Now, both owners are glad that I have the dogs on their farm. They feel like it helps keep off trespassers.

Our large male dog, Junior, literally acts like he is going to eat you alive when you startle him or approach the fence. All the hair is raised up on his neck and he comes charging at you barking viciously. Any sane person would never step foot in the same pasture with him unless they had a death wish. This is all a put on show. Junior will literally lick and slobber you to death once you talk to him, but he sure puts on a good bluff, which is exactly what you want.

We also had all the dogs fixed so they wouldn't come in heat. This would attract outside dogs. This is a cost that I am convinced pays dividends many times over. Male guard dogs that are not fixed will wander the country looking for female dogs. This is when guard dogs get run over, shot or stolen. The bigger and older they get the more expensive it is to have them fixed. Our local Humane Society has a special day each year that you can have your guard dogs fixed for less than half price.

We would rather buy the pups than try and raise them. We don't have the facilities or time to do this.

We have never separated the dogs from the sheep during lambing season. Some people will tell you to do this because they think the dogs' presence is too stressful on new mothers. I feel like this is the most vulnerable time for predators, and want our dogs with the sheep at this time for sure. Our older dogs have learned to keep their distance from ewes that have just lambed. Otherwise, they will get knocked to the ground.

We had a vicious predator of some kind last year kill our neighbor's Anatolian Shepard guard dog. Their farm borders our sheep farm. Whatever predator it was jumped into their pen of goats, killed one and ate the hind quarter. Their guard dog's throat was ripped out several hundred feet away. That very same night we had 82 yearling steers grazing across the road from this farm. Next day I moved them. Two 700 lb steers came up missing all of their tail except the last three inches.

There were no claw marks on the rumps of the steers, it just looked like you took an axe and chopped the tail off. I've never seen anything try and take down a 700 lb steer! Our sheep flock and guard dogs were one paddock over from the steers, and not one sheep was touched. Two guard dogs barking can sound like an army is waiting. This probably convinced the predator to go elsewhere. I'm still not sure what kind of predator was roaming the neighborhood. I did pack a rifle with me when moving steers the next couple of weeks though!

Do not be alarmed if you see your guard dogs lying around sleeping during the day. You may think that they are lazy and not doing their job. Believe me, they have probably been working all night patrolling and barking at distant threats echoed by coyotes or neighboring dogs.

Our sheep farm is about two miles from our house. I can walk out at night and hear our guard dogs barking from two miles away. It is just hard for a predator to think about killing when there is loud barking coming from the meal site.

When it starts getting dark, the guard dogs change into patrollers. What looked like a sleepy old lazy dog an hour ago has transformed into a security guard deluxe. If anything lets out a yip or howl our dogs are barking and may move in that direction warning the predator to stay away.

I can remember on one of our leased farms that the coyotes would literally come up by the barn and howl at me while I was doing chores. We now have sheep, cattle, goats, pigs and chickens at this farm with a guard dog. There is no more coyote yipping and howling behind the barn!

The guard dogs also keep the raccoons and possums out of our chickens. We have a neighbor who lives by one of our sheep farms. We call her "Goat Lady." She has milking goats and guard dogs as well. She loves to stop on the road beside the sheep farm at a good vantage point and watch the sheep. She gets out her binoculars and really studies them.

One day she stopped and saw Roo chewing on something. She got her binoculars up just in time to see the last of a raccoon foot go down her throat. I'm guessing the raccoon made the fatal mistake of coming up to steal her dog feed out of the self feeder. The coon became Roo's supper.

I wanted to see how our guard dogs acted with strangers approaching out in the field. I was concerned that they would bite somebody or possibly a landowner. So, I intentionally crept up on the flock and the dogs out in the middle of the pasture one afternoon. From about 100 yards distance I stood up beside a tree.

The dogs were sitting up, one on each side of a group of sheep watching over their surroundings. Roo spotted me first and started barking, joined by Junior. Both dogs had their tails stiffly set behind them, and the hair on their necks was standing straight out. They were lunging up in the air and viciously barking at me. All they knew was there was something strange in their pasture that did not belong there.

Talk about being intimidated. I didn't know whether to hold my ground, climb a tree or take off running. They came down the ridge at me and stopped about 50 yards out continuing to punish me with their loud barks and protective posturing. I held my ground and with the nicest voice I could muster called out to them, "Hey Roo, Hey Junior."

They still were not sure who I was and continued barking and circling back and forth in front of me. I never walked toward them. They finally got closer and recognized me. They both put their hackles down and came trotting up to me for a pat. I believe if I had been a stranger, I'm not sure I would have held my ground.

If the guard dog leaves the sheep flock and follows you around, that is a bad thing. I do not mind the guard dog coming up for a pat, but it should return to it's family (the sheep), not follow you around.

One final word on young guardian dogs. It takes most of them a year or more for them to develop into quality guardian dogs. They will make a lot of mistakes their first year. If you can be patient with them that first year there is a good chance that you can help them develop into a great guard dog.

The guarding ability they possess is all instinct and bred into them over the centuries. That first year you just have to help them make good decisions. When they do something good praise them with a "good boy."

When they do something bad you have to scold them immediately after the deed.

If you scold them later, they don't have a clue what you are scolding them for.

High Density Grazing

Chapter 22
An Overview of Holistic Planned Grazing

The term "Holistic" as used here means that we are managing for the health of everything. Holistic management focuses on the importance of working in sync with nature to mimic natural processes.

With Holistic Management we can restore biodiversity, build healthy soils, create clean plentiful water, improve wildlife habitat, produce healthy food, improve our quality of life, and restore communities. Every action and decision you make has an effect on everything in your operation. Holistic Management International of Albuquerque, New Mexico, has some very good books on all aspects of Holistic Management.

In 2006 I was fortunate to be at a meeting where Ian Mitchell-Innes was speaking in Colorado. Ian is a South African rancher who owns 14,000 acres and practices Holistic Planned Grazing. He is a Certified Holistic Educator as well.

Ann Adams from Holistic Management International helped clarify all of these terms: High Density Grazing (HDG) is one component of Holistic Planned Grazing (HPG). HPG, as taught by Holistic Management International focuses on using livestock to not only improve profit, but also land health and quality of life of the producer. It is one of the few Whole Farm/ Planning management tools that integrates grazing planning with land and financial planning as well as time management.

While High Density Grazing makes excellent use of stock density and herd effect (and is a term most people recognize), there is more to continued land health and animal performance than these management techniques. HDG graziers who are achieving consistently positive results on the land and with their livestock are actually Holistic Planned Grazing (HPG) graziers.

Ian talked about what he was doing in South Africa with Holistic, High Density, Planned Grazing. He is grazing 4000-6000 head of cattle on a 100-acre paddock for one day, then he moves them to the next 100-acre paddock to repeat the process. The total stocking rate is 4-6 million pounds on the 100-acre paddock. The results he is having with this system are breathtaking. I could hardly believe what I was hearing.

He has used this system for years. When he first started, his ranch was not very productive. Since switching to Holistic, High Density, Planned Grazing his ranch has exploded with all kinds of positive changes. Each year he grows more grass than the previous year. His problem now is buying enough cattle to eat the additional grass that his livestock grow each year.

He said that when you walk on his pastures it feels like you are walking on peat moss. The pastures are spongy feeling from all the humus and microbe activity going on below the soil surface.

I could not take it any more. I raised my hand and asked him, "How much lime and fertilizer did you have to put down to get your farm going in this direction?"

His answer was none! The mob of cattle are responsible for all the improvements.

This seemed too good to be true. New plant species were starting to appear due to the tillering of animal hooves from the mob. More plant material, (litter) is being trampled onto the ground, which feeds the microbes. Manure and urine distribution covers the ground. Dung beetle populations have exploded.

Ian's ranch has become home to over 18 species of large game. The wildlife that has moved onto his ranch are a direct result of the High Density Grazing with his cattle. The forages are so rich and diverse that it is like a magnet to wildlife. The highly microbial soil grows very nutrient dense plants that the wild animals crave. He has turned this abundance of wildlife into an additional profit enterprise by selling guided hunts. For every species that you can attract onto your land, you get eight additional ones. In other words, each species supports others and builds more diversity.

Ian is also predator friendly. By not shooting predators, he has a resident group of predators that live on his ranch that feed on wild game on the ranch and don't bother the livestock. If Ian shot every predator on his farm that he had a chance to kill, he would constantly be under invasion by new predators that might kill his cattle instead of wildlife. Ian calls it "Re-educating the new chap on the block." "New chap" means new predators.

Ian prefers to call himself a microbe manipulator, not a grass farmer. Without an effective microbial action in your soils you will never grow more quality grass. A good oxygen ethylene cycle promotes microbe explosion under the ground.

The biggest lie ever sold to farmers is purchased fertilizer. By putting down nitrates, you kill the microbes and are left with sterile soil. The fertilizer companies laugh all the way to the bank because they have their hook in you now. You become addicted to their fertilizer every year because your soils are dead, and will not grow anything without their chemical help.

Worming, vaccinating, and feeding have been eliminated from his operation. By selecting the animals that can perform under this management his costs have plummeted and his cattle are healthier. Ian uses ph paper to measure the cow urine. This will tell you what your cattle need. A cow's ph should be neutral, which is a 7. With too much protein, the animal's urine ph goes to 9 and your animal shuts down. We need to step back and let animals be animals.

Droughts are no longer an issue on his ranch with this type of management. Ian leased out 7000 acres of his ranch this year to his neighbors simply because his cattle could not eat all the grass. His son has come back to the ranch and started his own herd of 500 cows along with Ian's 4000 cows! His creeks and rivers, which were dry have started flowing again due to the increased catchment of rainwater that falls on his humus-enriched soils.

Combining more cattle together is the cheapest thing you can do to get the biggest bang for your buck. Until you do, this you will never be able to achieve the above results.

Needless to say, I was overwhelmed with all the wonderful results Ian had, just by increasing his herd density along with a complete understanding of Holistic Planned Grazing.

When his talk was over, I hung around in the shadows and listened to Ian answer additional questions. That night back at my motel room I tossed and turned all night - could not sleep a wink. All those wonderful results that Ian reported were attainable with no added costly outside inputs!

The last thing Ian told me was to concentrate on building the mob and ignore everything else. The mob will make you more money than anything else you can do. You must harvest free energy dollars to be sustainable.

Well, after a sleepless night I was more fired up than ever. I immediately started reading everything I could get my hands on about Holistic, High Density, Planned Grazing. I signed up for a Holistic Management School that winter. In the time leading up to the school I read Allan Savory's Holistic Management book, *A New Framework For Decision Making.* It was one of the toughest books I've ever read, but I fought my way through it. It was just an awesome book, packed with all the Holistic practices Ian had talked about and a lot more. The Holistic School was offered by Kirk Gadzia in Albuquerque, New Mexico. It was six days of very good information.

I came home ready to take on the world with all this new learned information.

Chapter 23
Holistic High Density Planned Grazing with Small Herds

When I first heard of Holistic High Density Planned Grazing, I assumed you had to have a large number of animals to perform this practice.

I will admit, the larger the herd the better and more impressive the results will be. I can never imagine matching Ian's 4000 head cow herd, but I have had some very impressive results with smaller cow herds.

We have been practicing High Density Grazing for almost two years on our grazing operation with cow herds that ranged from 50 to 250 head.

No matter how small your herd is, you can still produce amazing results by using High Density Planned Grazing. We have seen some very positive results with mobbing up our cow herds to densities of 100,000 to 500,000 lbs liveweight/acre. This would be equal to 100 to 500 cows per acre. We have never seen anything in the grazing world have such a dramatic ecological, financial, and social effect on our farms.

We have not tried High Density Grazing with sheep or goats yet. The advantage to using cows with High Density Grazing is that they are a heavy animal. You have a bigger work tool, which allows maximum soil disturbance, trampling of litter, incorporation of urine and manure into the soil surface.

With our previous Management-intensive Grazing system we were doing fairly well, I thought. As a matter of a fact, I thought it couldn't get any better than that.

I was hugely mistaken.

With MiG, we basically moved livestock every two days, except for the dry cows, which we moved every three days. Under our old grazing system, with two to three day moves, rest periods ranged from 20 to 60 days, depending on the growing season and moisture. The paddocks would be grazed fairly evenly where the grass was good. In other areas of the paddock where the grass was not so good, the grass was barely touched. We could get cattle to eat some weeds when they were young and tender, but once mature they were completely ignored.

One very important practice that we were missing with our previous MiG grazing system was not allowing the plant to fully re-grow before grazing it again. We were concentrating very hard on keeping the plant young and vegetative to keep seed heads from forming. The livestock ate those plants well. The problem we were unaware of was the roots were not fully re-grown.

As the growing season progressed, the forages stopped growing when weather conditions turned hot and dry. We were always thankful to ship stockers or cows in the middle of July because our grass was not growing back due to our stunting the roots in the early spring by not allowing them to fully re-grow.

Look at it this way, when the young plant is starting its growth in the early spring and you bite its head off before it has a chance to fully put down roots, you have forced the plant to start re-growth with a half tank of fuel in its roots.

With our new High Density Planned Grazing system, depending on what part of the growing season we are in, herds are moved from one to three times per day. Our rest periods between each grazing cycle ranges from 25 to 150 days. The plant is fully rested along with the roots fully re-grown.

This rested grass sward is a beast. It can handle just about any weather condition or grazing treatment it encounters. It's amazing how fast a fully rested plant grows back with all that stored root energy.

We now have a system that puts pressure on all plants

in the paddock evenly. Nothing escapes animal hooves or mouths.

By bunching up the livestock there is simply more competition to eat what is offered. We try to get 60% consumed, 20% trampled on the ground, and 20% left standing. The actual results may vary some from this ideal ratio. Sometimes the livestock take 80% and trample the remainder. This is not a problem as long as you let the paddock fully re-grow before exposing the livestock to it again.

We have changed our attitude toward weeds and brush. They are no longer a nuisance. Both are livestock feed with High Density Grazing. With our leased farms and not owning a tractor, weed and brush encroachment had been a management concern.

Our dry cow mob is our best tool for this. Their nutritional requirements are not as high as a lactating cow, so we simply bunch them up, and they eat and stomp everything.

We have a period in July-August when ironweed and lanceleaf ragweed simply take over. We've never had much luck getting cows to eat it. This is no longer true with the dry cow mob. I will go into more detail in a later chapter on some of the methods we use to graze weeds.

The energy that the mob injects into the soil is absolutely amazing with each grazed strip. The high density of cattle hooves press the fresh manure into the soil surface, which is like giving the grass a booster shot of adrenaline. The paddocks look like you took a floor broom and swept them with manure and urine with the higher stocking densities.

One thing I have noticed at these high stocking densities is how well the grass gets watered with urine from the mob of cows. It is like giving the whole paddock a drink of manure tea, great stuff for a plant! This extra moisture dropped on the plants from the cow is huge when you are in the middle of a drought period. We monitor the re-growth on paddocks where the cattle have been removed, and the grass has re-growth within 24 hours if moisture is present.

Our method for monitoring re-growth is something I picked up from the Holistic School. When the cattle are removed from the paddock, we bend a wire flag over to the height of the grazed forage. We then go back periodically and measure how fast our grass is re-growing. This gives you a very good view of how fast to move your herd so that you do not go back to that paddock until it is fully re-grown.

Normally grass does not start re-growing for three days under conventional grazing. This is not true with High Density Grazing. We see immediate re-growth in 24 hours.

I have a picture that I took in the spring rotation that shows the cows in their paddock with the previous day's paddock in the background. The previously grazed paddock has at least one inch of bright new green leaf sticking up - within 24 hours! That is why we always keep a back fence to prevent the cows from going back and eating off the previously grazed plants. They definitely will go back and graze off that precious new inch of new growth. This just kills your future growth on that paddock. Do your grass a favor, keep in that back fence.

We are using the same grazing paddocks that we used with the previous system, just breaking them up into numerous temporary paddocks to get our animal density higher.

We are also using the existing water systems with some modifications, which I will explain in a later chapter.

Temporary lanes are used when needed to get the cattle to water with no adverse effects on the land or cattle. Manure deposited in the lane is actually being placed on the pasture, not in a permanent lane. There are no permanent dirt trails simply because the animals are not on the lane long enough for trails to develop. The temporary lanes are moved to a different location with each grazing cycle.

We have developed fast effective grazing procedures for putting in temporary paddocks and lanes, which will be covered in detail in a following chapter as well.

We started out with two mobs that we grazed. One group is made up of 140 dry custom grazed cows, the other group is our owned cow herd along with 70 dry custom grazed cows to form the mob. Custom grazing cows has allowed us to build a mob by using other people's cattle. We are using last year's high density grazed paddocks on one farm to strip graze our grassfed yearlings this year with good results.

The positive results we have seen include improved quality of forages, more forage quantity, more plant diversity, better manure distribution, much more drought proof paddocks, tremendous microbial action, increased water retention, better mineral cycling, perennial grasses coming back, more dung beetle activity, more diversity of wildlife, higher stocking rate over time, and more profit.

All of these awesome results have been achieved without any major costly outside inputs. The only investment made was geared ratio polywire reels and some tread-in posts.

The picture below shows the grazing effect of a half day move with 300,000 lbs stocking density.

Chapter 24
Our High Density Grazing Fencing Techniques

Our High Density Grazing techniques will work fine for any grazing system that is using some type of rotational grazing. It is easier to explain by showing, but I will try to verbally describe our grazing technique.

All of our farms were previously set up for rotational grazing. Most have paddocks that range from five acres up to 70 acres in size. Since we started High Density Grazing we have not added any more hi-tensile wire paddocks. We are using what was previously there with no problems.

Most of our paddocks have hot wire on at least two sides. If there was a good perimeter barb or woven wire fence in place, we left it when we constructed our paddocks. So far it has not been a problem because there is a power source in every paddock where the water is located.

We always start at the water source and work our temporary paddocks away from it.

The first temporary fence of polywire we put in splits the grazing area in two pieces.

Visually walk or drive your ATV across the center of the paddock starting at the water source.

We put up all of our polywire cold (no voltage). Some brave folks run it hot.

We tie the wire around the fiberglass post with a simple slip knot that leaves a loop sticking out. That aids in removing the knot when we take it down.

Let's say you have a barb wire or woven wire fence on

one side of a paddock. All you have to do is take one of the white polyethylene 42" O'Briens step-in posts and stick it in the fence line on the back side. Step it securely in the ground down to the spike level.

Now the side of the fence that you are standing on, which is the direction the fence is going to run, is where you reach down and tie the polywire onto an open area of the polyethylene post that is not going to touch the barb or woven wire when you tie your slip knot.

Are you confused yet?

All you are doing is using the conductive fence as a brace for holding the white polyethylene post secure. You could also use a red O'Briens pigtail post. Just step it in on the same side of the fence that you are standing, tie on and start rolling out your fence. Sometimes I don't have a red pigtail post with me. That is why I came up with the permanent fence method for securing a nice solid non-conductive corner post.

Now we will string the wire across the paddock putting in step-in posts every 60-100' depending on the terrain of the paddock. If you have multiple dips and hills, the posts need to be closer together to keep the wire at the correct height. For cows, 32-34" is about right. With yearlings, 28-30" height works real well.

As we make our way back to the water source, your ending target should be to the front or back of the water source. With the O'Briens reel in your hand, held over the fence that is hot, pull all the slack out of the polywire that you have just unrolled across the paddock by rolling it onto the reel.

Now take the cold polywire, and with your free hand wrap the wire around the O'Briens hook three times. Set the hook on your powered hot wire. This powers your hot wire without using a handle that you have to pack with you and tie on. Once you do this the entire reel is hot except for the insulated rubber handle. It took a year of tying handles on for the power source before I figured out the hook on the reel would do the job just as well, and I didn't need any handles!

165

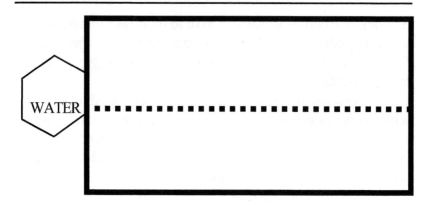

The dotted line above represents the first temporary paddock division installed. Always hang the reel at the water site, which gives you maximum flexibility.

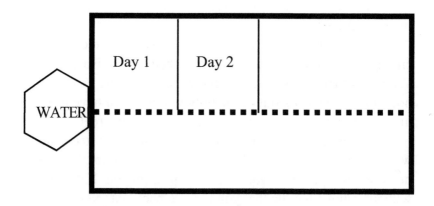

The dotted line above represents the first temporary paddock division with two paddocks fenced for grazing days 1 and 2.

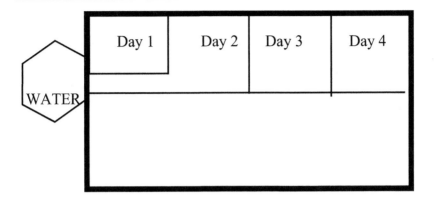

On day 2 the lane is extended for the livestock to reach the watering site. The back fence is left in from day 1 to prevent backgrazing.

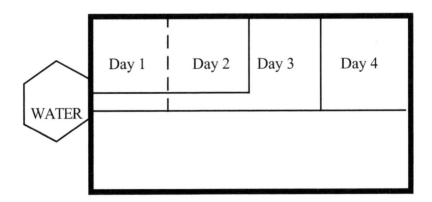

On day 3, the land is extended to the water site with fencing to prevent backgrazing in paddocks 1 and 2. The dotted line represents where the previous paddock division was.

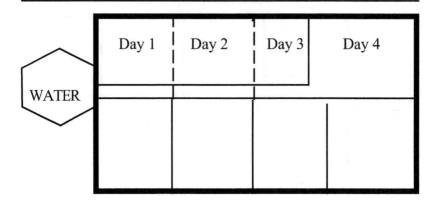

By day 4 the temporary lane is extended to the water tank past the previous days' grazing. The dotted line represents where the previous paddocks were. Days 5 through 8 are installed, ready to graze.

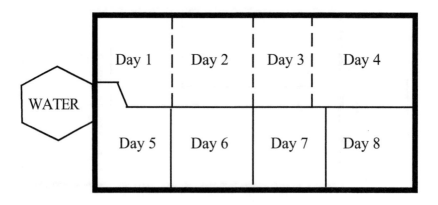

This sketch shows the reel moved to the other side of the water point in order to service days 5, 6, 7 and 8. Extend the temporary lanes just as done in the previous days of 1 to 4 divisions.

Okay, now we have split the paddock into two semi-even pieces. We are ready to put in our first temporary paddock.

Let's assume that we have a herd of cows that we want to move morning and night. So we need to calculate how big of an area to give them for half a day so that we get the correct graze/trample ratio that we want. The Holistic School taught us that 60% consumed, 20% trampled, 20% left standing is the ideal ratio. Our results vary from day to day. Sometimes we consume more and trample more, which leaves less standing.

The following is only an example. The actual footage numbers will vary with each farm location and pasture. Some pastures are just better than others.

The way to do this is by asking yourself, "If I measure off an area say 30' x 30', will that feed a cow for half a day?" Whatever the footage answer is, multiply that number by the number of cows that you have in your mob. All you are doing is taking an educated guess as to how much the cow will eat.

Walk off the footage that you estimated, and this is where your first temporary paddock will go. You will know within 12 hours how good of a guess you made.

Adjust your next paddock based on the results you had by your first guess. If they strip it clean, you under estimated how much a cow would need for half a day. If they partially grazed it and you have lots of standing forage that is not trampled, then you over estimated how much a cow would eat in half a day.

You really develop a good eye fairly quickly as to how much to give the mob after several moves. You don't have to accurately measure every paddock installation that you put up. You may want to measure each paddock until you get your confidence built up over time.

If you guess wrong one day and they severely overgraze an area, it is not the end of the world. With High Density Grazing, you are building such a long rest period into your grazing system that you will not be grazing that paddock again until it is fully re-grown. I don't lose any sleep over it anymore.

Let's get back to the water tank and continue on with this mob that we just built our first paddock for. Your first paddock can be any shape you want. All pastures are not perfectly square, so don't worry about it. The more square they are, the more even your graze will be. Make sure they can get back to the water. Most of our paddocks end up being rectangles because we usually can not move the water source.

Okay, before you turn your mob onto the first paddock, always put up the second paddock that they will be turned into next. The reason for this is that if the mob would accidentally knock down your temporary wire they will not have access to the rest of the un-grazed field.

If you see that you did not estimate right with the first paddock, you can always move the wire in or over before letting the mob into the second paddock.

When putting in this second paddock, make sure that you end up with the actual reel at the end that is closest to your water source. This reel will get its power from the polywire fence that you originally put up to split the whole paddock in two.

There are two ways you can power this reel.

One way is to wrap the polywire around the O'Briens hook and lay it over the hot wire at a post location.

The other way is to tie an electric fence handle on the polywire and hook it to the hot wire at a post junction.

The temporary division fence is not strong enough to hold up the weight of the reel. That is why we use these two methods.

When I have spare time, I usually put up four to five paddocks in front of the mob ready to graze. This gives you a cushion if you get in a time crunch any particular day.

When you go out to move the mob to their second temporary paddock, unhook the handle or pick up the reel that is laid over the fence (whichever method you used) and wind the reel back about 50 feet to let the mob through to their next paddock. Don't dilly dally around when you do this. The mob is anxious to get to the fresh grass, and you need to get out of their way.

Once all the cattle move through to the fresh paddock, walk back and stick in a red O'Briens pigtail post about 16 feet (the lane width will vary depending on mob size) from the hot division fence that you unhooked it from.

Position the pigtail so that the loop is holding the wire in the closed part of the pigtail.

Next walk back toward the water source extending the lane to the back of the water source.

Hook the reel on the hot powered fence and you have a water source for your mob. You also automatically have a back fence erected, which keeps them from over grazing the paddock that they were just on.

I used to get lazy, and not put in a back fence or lane. I just let them graze forward with each paddock change. This was a huge mistake, and I got burned. My excuse was, "Heck, those cows will not graze that old paddock when they have the fresh one that they are on."

What happened was the cows would graze the fresh piece first, then if they were short or just bored they would go graze the previous paddocks shorter. All of this was happening when the grass was trying to re-grow, which is the most vulnerable time of a young plant. This will set your pastures' growth rate way back because the baby plants get their heads chopped off. Then they have to use their short little roots to start all over again.

Several other things were happening when we did not back fence. Our manure distribution was not nearly as good because the mob had no pressure put on them by having all the open area behind them.

They may even go back and lay down in their favorite lounging area, which concentrates all the manure in those areas. With the back fence in, their manure and urine will be very concentrated in the new paddock.

One other thing we were missing out on by not back fencing was the trampling effect of the mob. By not being mobbed up, the hoof traffic was not intense enough to trample the forage.

Also we noticed the selectivity of forages by the cattle increased with no back fence pressure. They left weeds and other less palatable areas untouched.

The time it takes to extend the lane with each paddock move is well worth it from the results you get from having it fenced off. I know it is hard to believe the results will be that much greater by keeping a back fence erected, but it is the truth.

We all have heard a lot of negative comments about lanes in our grazing systems, but we are not talking about a permanent lane here. This temporary lane is only used for grazing that first quadrant that we split in half. The average time that we are on a temporary lane is four to five days, then the lane gets rested for 70-120 days, depending on the growing season. The next time you come back to graze this particular rested paddock, the lane area is not even visible. The re-growth in the lane will be higher quality than the strips that you grazed. The lane saw more animal traffic along with lots of manure and urine deposited the full length of it.

By the fifth day in most instances, the mob is just starting to get a path beat down to the soil in the lane. If you use a temporary lane more than five days, you can expect to get a dirt path exposed.

As you extend your lane each day to provide a water source to the new paddock, you only have to add a post or two to your lane. When you get all the first half of your original paddock grazed, you are now ready to graze the other half with your mob. Have your first paddock in by the water tank, then unhook the original first reel that you used to split the paddock in two and reel the wire back to let the mob enter. Once the mob is grazing on the fresh paddock, take the reel and walk it over to the back of the water source. Hook it onto the hot power source and you are done.

Now the mob is positioned to graze the other half of the paddock that you originally split in two. You can also remove the lane from the other previously grazed side. You will repeat the same process on this side as you did the previous paddock that you just came from. Your new lane will be parallel with the one you just took down. One side of your lane will always be there (the wire that split the paddock in two when you started).

Look at the illustrations on the previous pages in this chapter for a visual description of what the paddock will look like.

It's simple to put in, simple to take out and move to the next paddock.

One thing I will note here is that the mob is so concentrated on grazing their fresh new paddock, that you can leisurely take your time putting in the new lane, removing wire, etc. They could care less about what you are doing behind or beside them. This makes it nice when you're limited on reels and you have to gather a reel that was used in the old lane to reuse for the new paddock.

Congratulations, you made it through fencing your first high density grazed paddock.

Let's assume we were on this paddock for five days on each split side. That's ten days total grazing on the paddock.

Walk back and look at your first temporary paddock that your animals grazed. If moisture is present, you probably will have some pretty good re-growth started back already. If you left out the back fence you will not have any re-growth.

I have a spring growing season picture on one of my speaking presentations that I give at various grazing schools. It shows a paddock in the background that was mob grazed. In the foreground is the mob on their next paddock adjacent to it, one day later. Within 24 hours, you can see re-growth starting on the paddock that they were on the day before. Bright green growth is exploding back into the pasture with the mob removed one day. It is hard to argue against the importance of

173

keeping in a back fence after looking at this picture. I had no idea that grass would grow back that quickly when we started mobbing up the cattle.

We are finding that where we previously got two days on a paddock with our old grazing system, we are now getting at least six days or more with the high density system When you multiply that times the number of paddocks in your grazing system it really makes a big difference in your rest periods. We went from grazing a paddock six times per year with MiG to two to three times a year with High Density Planned Grazing.

There are several more grazing tips that I want to go over with you that we have learned over the last couple years using High Density Grazing.

If you have a tractor, sell it and buy an ATV. The ATV will literally save you hours per day putting up temporary wire. We would absolutely be in a world of hurt without our ATV.

Yes, it is a Honda, the most durable and reliable ATV that is made. We have had others. None of them was on the same playing field with Honda.

We have a one and a half inch pipe bracket that an O'Briens' reel handle will slide down into that is bolted onto each side of the rear cargo rack. The pipe depth is 6" with a bottom welded to it. This prevents the reel from going too deep into the pipe, which allows the reel to freely spin without the hubs hitting the pipe. The pipe bottom is a piece of flat steel that is 3" wide by 10" long. This flat steel piece extends over each corner of the cargo rack and is held in place by one U bolt. The way you get around just using one U bolt per reel holder is that the flat steel piece goes over the first rear bar of the cargo rack and then slides under the next steel bar on your cargo rack. I have probably confused you.

Let me try a little different approach. Visualize trying to pry a huge boulder out of the ground with a pry bar. To do this you will need to set a rock or something to lay your bar on to pry against. This rock that you just set there to pry against is the rear steel bar of your cargo holder. This is also the bar that will have two holes drilled in it to secure your U bolt to.

The boulder that you are trying to pry out of the ground is the next steel cargo bar that your flat steel end will set under. This bracket frees your hands up when unrolling the polywire.

The reason I have two brackets is that I can unroll two wires at once if I need to. The brackets have about a 30 degree angle away from the back of the ATV, which gives the polywire a perfect angle to unroll.

The one and a half inch pipe is clamped onto the cargo rack where they actually extend out past the back of the 4-wheeler. The advantage of them sticking out past the back of the ATV is that they are not in the way of packing items on your cargo rack.

When we roll up polywire, we never walk the fence pulling posts and reeling polywire at the same time. There are two reasons for this.

175

The first reason is that when you walk along and roll up polywire it is impossible to put any tension on it. By the time you get all the polywire wound on without any tension, it is a floppy, loose mess.

The real problem comes when you try and unroll a floppy loose reel. You will be cruising along and a huge loop will appear and jump off the reel around your spindle or the hook. By the time you notice it, your polywire is broken.

The second reason we do not walk along pulling posts and reeling up polywire is that we do not have to pack the reel or posts. By cruising along the fence with the ATV we just reach over and put the post on our cargo rack post holder. There is no reel to mess with. You can just concentrate on pulling posts.

We pull the posts first, then go back and roll up the polywire by standing and turning the reel as fast as it will go. I have not met a person who can keep up with my polywire end piece as I am rolling it in. This reel is lightning fast!

I was running short on reels the other day and regrettably used one of my old 1:1 ratio reels. Luckily I only had about 1000 feet of polywire to roll up. Man, I thought I would never get to the end of that wire. My arm got tired reeling the crank, and the wire seemed to be just creeping along the ground. Okay, I admit it. I am spoiled to good reels.

Seriously folks, you're going to be doing a lot of temporary fencing. Make it easy on yourself with good geared reels.

I am always looking for a better or more efficient way of doing a grazing task. This method I am about to explain takes a lot of the work out of rolling up long lengths of polywire that you have unrolled when erecting a temporary paddock.

Let's say you have 2000 feet of polywire strung out across the paddock in which you split it in two. Even with a geared reel, this 2000 feet is going to be tough to reel up if you stand in one spot. The reason for this is that lengths of 2000 feet put a lot of ground friction on the string. Granted when you get 1000 feet of it on the reel it will get easier to crank. But that first 1000 feet will eat your lunch.

Here is what you do. Position your front ATV tire right beside the polywire in the direction that you want to roll the wire up. Grab the reel and place it on your lap or set it in the reel holder if you have one on your ATV.

Drive along the polywire with your ATV until you get to the center of the string (1000'). Now get off and roll up the polywire. You will be amazed how easy the string is to roll up by taking the full 2000' of weight and ground friction off of it. The string is actually doubled back on itself, so the first 1000' is not pulling the other 1000' when you roll it up.

It is kind of like rolling up two 1000' lengths, much easier than tackling 2000' being pulled the full length from one end.

This simple little trick will save your arm, and the time it takes to roll it up is cut in half. This is because you can turn the crank so much faster with the reduced weight off the string. It took me a year to figure that one out!

Another fence building shortcut we use requires two people, but cuts the time to put up a fence in half. This technique would be great to use if you have access to some younger children. I have an older retired fellow who is slightly overweight. He helps me put up fence some nights. Working with me took 30 pounds off of him one summer. He told me that he feels better and has better wind than ever. It gets him out of the house, he receives his exercise, and gets a little spending money to boot.

Here is our two person method.

All the posts are secured on the cargo holder of the ATV. The reel is in its holder. Both of us get on the ATV and drive across to the end of the temporary paddock where we are going to tie on the polywire cold. My neighbor gets off and I take off across the paddock spearing the tread-ins into the ground right where I need them, kind of like throwing lawn darts.

I visually estimate how much grass to allot the livestock based on the quality and volume of forage. I also identify the low spots, humps and curves, which is where extra posts are needed. My neighbor does not have to think about where the wire or a post goes. He does not have to pack a post or reel. He just grabs the post, steps it in, and hooks the wire on the hook. By the time I get to the end of the temporary paddock where I want to hang the reel to power it, he has almost caught up with me.

One note here. Do not hook the reel onto the powered fence until your partner has the last tread-in secured! That may be a good way to lose your help.

In thirty minutes, my neighbor or my wife and I can have a lot of temporary paddocks put up. The same procedure works for taking them down. I drive my helper to the end that is cold and then drive over to the power source and unhook the reel off the fence. I wave my hat to show him when it is cold. He unties the cold wire and gathers the temporary paddock posts while I am taking up the lane posts.

Chapter 25
Our High Density Grazing Water System

When we switched to High Density Grazing in 2006, we made some minor changes to our watering systems.

The first thing we did on all of our gravity flow water systems was to switch from 3/4" water tank valves to 1 1/2" brass tank valves. The larger 1 1/2" valve has been the salvation for watering large herds of cattle. When you have large numbers of cattle come to drink at one time, you must have either a large water valve for fast refill or a very large water tank. The brass water valves are very durable and will last your whole lifetime of grazing.

Our old 3/4" gravity flow valves hooked to 400 gallon tanks worked fine for herds that numbered 80-100 head. With herds of 150-250 cattle, the tanks could not keep up with the smaller valves. The 1 1/2" brass valves are more expensive, but they are worth every penny. The actual float that sits on top of the tank that controls the level of the tank is made by Jobe. It is a bright yellow very durable polyethylene ball float that is impossible for livestock to tear up. I love them because we have never had one break or crack. The actual brass valve that we use is made by a company called Dean Bennett Supply Company in Denver, Colorado.

The first water tanks we purchased were round galvanized steel tanks that held 400 gallons. These tanks are still in use but we use them on pressurized water systems. We have several 600-gallon polyethylene water tanks that have held up very well, but they are expensive.

The latest tanks we bought have been the best so far. They are large earthmover tire tanks. These tanks cost us 10 dollars each and are very durable. You basically can not tear one up. They do not rust, rot or crack and are bullet proof. The rubber is so thick that you have to really work at it to cut a slice out of it.

Our tire tanks hold 800 gallons and have a 1 1/2" brass valve plumbed into the bottom of them. I love these tire tanks and plan to put in several more of them where we have ponds to supply them.

We got our tires from a tire recycler that had truckloads of them. The tires were free, but they charged 10 dollars to load each tire. The tires weighed 2000 lbs each, so make sure you have access to a front end loader to unload them. We cut the sidewall out of one side of the tire. This allows the livestock easy access to drink from them.

The side of the tire that is placed on the ground still has the sidewall in it. We dug a trench for the 1 1/2" water line feeding the tire tank and also a 3" drain line exiting the bottom of the tire tank.

After the water and drain lines were trenched in, we tamped the trench area where the tank was going to set. A good tamping job keeps the trench from settling after you have poured concrete in the bottom of the tank.

Before we poured the Ready-mix in the bottom of the tire, we put down polyethylene geo-textile fabric around the outside of the tire. This fabric was covered with 12" of rock that extended out from the tank for six feet. This gives a very durable platform for livestock to stand on while drinking.

We placed 12 inch diameter cedar logs around the outside of the rock to keep it in place around the tire tank. These cedar logs were staked and wired in place. Rock is too expensive to not go through this extra effort of holding it in place. A couple hundred cattle drinking multiple times each day can move a lot of rock if it is not held in place by timbers.

By laying down the rock first, it holds the tire in place

while pouring the Ready-mix in the bottom of the tire. Also all grading work with machinery is already done, so you do not have to worry about bumping the tire and breaking the Ready-mix seal in the bottom. It takes about six to seven bags of Ready-mix to form the bottom of the tire tank. We mix it up in a tub and make it very runny. By having the Ready-mix runny, it allows cement to run back under the sidewall that is laying on the ground.

We pour in Ready-mix until it reaches the top of the sidewall, which usually gives you about a 4" thick bottom on your tank.

Hook up your water valve to your water line and place a cover on your drain, and your tank is finished. We do put several 2 1/2" fiberglass posts across the tank to keep livestock out of it. Wood posts would work fine too.

Another method we use to keep cattle from entering our water tanks is to run an electric polywire over the center of the tank. Our old boar hog climbed into one of our big water tanks to cool off. He reached up and touched the polywire with his nose while he was soaking in the tank. I have never seen an animal exit a water tank any faster than that hog did!

Another method for getting large herds to water is to use temporary travel lanes and let them walk to the water source. We have done this for several years with good results. These temporary lanes are hooked to your daily strip grazed paddocks. The cattle do not spend much time in the water lane. The fresh grass strip that you exposed them to is where you will find them. All day long you will have several cows walking to water, several coming back from water, and several at the tank drinking. We still got very good high density results by using these temporary water lanes. Do not bankrupt yourself by placing water sites all over the farm, let the animals walk.

Several of our farms have pressurized water systems installed. Pressurized water is a nice luxury and can really come in handy. We can still use 3/4" water valves with pressurized water to supply large mobs of cattle. We keep the water as close as possible to the mob on these pressurized systems by hooking a 100 foot piece of 1" black polyethylene 160 PSI water pipe to our permanent water points. This pipe has a 100 gallon tank hooked to it and is placed out in the paddock with the cattle.

We run a hot polywire right over the top of the water pipe and water tank. This keeps the cattle from stomping on the water pipe, and it also protects the water tank from the cows tipping or pulling on it. We love the flexibility of being able to move the water point around with each grazing to a new location. This is a very powerful tool for concentrating manure right where you need it as well.

We are going to build a tank that is mounted on a trailer that can be pulled with our ATV. This will allow us to transport the water tank full, instead of having to empty it when we move to the next paddock.

Chapter 26
Added benefits of High Density Planned Grazing

Now that you know how to build fence for High Density Planned Grazing, let's talk about the added benefits of this system.

Increased Stocking Rate:

The first year we used High Density Planned Grazing, we did not increase the stocking rate. We increased the density of our animals where they were being grazed, and this brought more grass.

The biggest mistake you can make with High Density Grazing is to increase your stocking rate the first year.

In our second year, we increased our stocking rate based on what we saw the first year. There were some farms from the previous year where we were severely under-stocked and we only got them grazed once.

I was a little nervous our first year that this increased animal density on our pastures might not allow them to re-grow at a rate that would allow them to be grazed when I needed them. On one particular farm, we went from 60 cows the first year to 134 cows the second year, and still only half of the paddocks were grazed twice before the cattle owners needed them back for fall calving! It sure was a warm, cozy feeling to have all that grass left over while were experiencing a long dry period.

One of the problems Ian is having with the High Density Grazing now that he has been doing it for numerous years is that the cattle keep growing more grass each year. He is having trouble buying enough cattle to eat all of his grass. I think that would be a nice problem to have!

We are monitoring our grass and gradually increasing the stocking rate as we go forward. This is the part that really gets me fired up. Can you imagine growing more grass each successive year just by increasing your herd density? There are no outside costly inputs. We're just being rewarded for good management.

Folks, this is the way to grow your operation and build a sustainable business.

Better Manure Distribution:

When you get up around 300,000-500,000 lbs /acre, (300 to 500 cows) the manure distribution on the paddock is unbelievable. It is hard to find a manure pile that has not had a foot or two dragged through it. Some will be flattened like you hit them with a floor broom.

It still amazes me how fast the manure piles disappear once the cattle have been removed. Dung beetles invade the manure immediately.

With our old MiG grazing management I thought we had pretty good manure distribution. Our old system is not even on the same playing field as our new High Density Planned Grazing system when it comes to manure comparison. I promise you cannot walk across our grazed strip without stepping in a lot of cow manure. The field has a very distinct smell to it, pleasant if you like the smell of money.

I get a kick out of bringing someone out to show them our grazed strips. If they are squeamish about stepping in poop, they never get their eyes off the ground because they are trying to dodge the mine field. I don't worry about it, I walk through it and spread it just like the cows do.

I talked a friend into trying mob grazing. He was using his uncle's John Deere Gator to move the wire every day. When he returned to the house each day he would take a garden hose and wash all the cow manure off the Gator.

His uncle asked him one night, "How in the world are you getting all that cow manure on my Gator?"

My friend asked him to come with him the next night to move the wire. The uncle volunteered to drive the Gator while my friend put out the step-in posts. My friend looked up after his uncle had driven several feet and his uncle exclaimed, "You can't drive through this field without hitting a manure pile!" His uncle looked like he was going through an obstacle course trying to miss every cow pie. At first his uncle thought he was nuts when he started the mob grazing, but now thinks differently after seeing the results.

By giving the mob small areas, they stay on the new strip almost entirely because that is where the fresh feed is. You will get some manure around the water tank, but the majority of it is spread around the new strip of exposed grass. By moving the herd daily - more moves are better - the flies are always on the fresh manure left behind by the mob.

We had several area cattlemen come by for a farm tour in July 2007, the peak of fly season. They kept asking where were all the flies? The cattle on their farms were loaded with them, and they could not understand why ours were not. It is always a good thing to be moving onto fresh undisturbed grass, leaving the majority of flies and manure behind.

Increased Quality of Forages:

Our overall forage base has improved in quality with the High Density Planned Grazing system. We previously had areas on every paddock that livestock refused to eat. After the first rotation of the growing season these rank, stale forage areas were avoided all season. If you miss giving an area in the paddock animal impact with your first rotation, you have missed an opportunity to increase the productivity of that paddock. With High Density Grazing, the animal impact on these areas makes the forage quality skyrocket.

I have never seen anything like it. The mob effect just

brings quality to the forefront. Also, the wildlife is really starting to be drawn in by the succulent salad bar pastures.

It is amazing how long the plants hold their quality with this system as well. I have never seen mature plants hold onto their quality so long and not put up seed heads! You can count the seed heads. Most of your re-growth is succulent leaves. The more High Density Grazing treatments these plants see, the better the quality. A healthy sward has less seed heads because it is not worried about survival. If you have control over a piece of ground, you should focus on getting as much quality grass growing on it as possible. High Density Grazing will do this for you.

More Quantity:

With all the added manure distribution, we are growing more forage than previously. The rested, fully re-grown paddocks look like you put 100 pounds of nitrogen on them, and the plants have very dark green colored leaves.

My neighbor asked me how much fertilizer I had put down.

I replied, "A whole mob's worth."

The long rest periods allow all your plants to fully re-grow. This rested plant is a strong plant and can withstand severe droughts better than half grown plants. The mature sward also shields the canopy from the sun and helps preserve precious soil moisture loss, which keeps the plant greener.

Our fully re-grown rested paddocks are hard to walk through now. The plant spacing is very tight, with less bare ground. With the price of fertilizer skyrocketing, it is very satisfying to be able to grow all this extra grass just by changing management styles.

In the spring of 2007, nitrogen fertilizer hit 58 cents per pound. That comes to 58 dollars per acre if you spread 100 lbs. With High Density Grazing you can keep that 58 dollars in your pocket to buy more livestock or whatever. You'll have more quantity without any purchased expensive fertilizer. Now that is something to get excited about. After all, the money to be made in this business is from the grass. Look at the livestock as a tool to harvest the money.

The more we can stimulate the soil and grow microbes, the more money we can harvest. If you put down chemical nitrogen on your soil, you will have little microbial activity. The nitrogen kills the living microbes in your soils, which leaves you with sterile soil that must have its annual dose of purchased fertilizer to grow anything.

Increased Plant Diversity:

Grass species that we have not seen before are starting to appear in our paddocks. None of these new species were seeded, but resting in the seed bank waiting for the right opportunity to grow.

I believe there is no better way to get diversity of forages than to beat it up with a mob, and give it a nice rest to express itself.

Redtop has exploded in some of our paddocks. We never had any before. Several years ago I priced redtop seed because I was interested in getting it started on our farms. The price at that time was $4.00 per pound, and the seeding rate for redtop is three to six lbs per acre. If you took the top seeding rate of six lbs, that is 24 dollars per acre saved in seed costs just by mob grazing, plus the labor of spreading the seed. Redtop seed count is 4,900,000 seeds per pound. That is a lot of seeds.

There are some areas on our farms where redtop is competing very well with what was a stand of rank fescue. The reason is, we trampled and grazed it severely for half a day, then allowed it whatever time it took to fully recover. You will not recognize the same pasture 60-90 days after beating it up with the mob. It will look like someone waved a magic wand. The long blades of grass also collect massive amounts of dew from all the surface area of the large leaves. This moisture runs down the leaf to the stem, which guides it into the roots. I went out in the pasture one hot afternoon with a fellow who I was giving a farm tour. I pulled back the huge green canopy and took my pocketknife to cut a wedge out of the soil about three inches deep. Although it was 104 degrees F that afternoon, the bottom of that hole was cool. This was the temperature that the grass plant roots were exposed to as well. No wonder the forage was green and in very good condition with no rain after a month of extreme heat.

We never had experienced that type of grass persistence in that type of heat. Our pastures always went dormant and brown when it got hot and dry.

I have never had a summer where I rested so well knowing that I had plenty of forage and never had to de-stock anything, when it didn't rain. You will never be put out of business, because you will always have grass to graze. We are only in our second year of this High Density Planned Grazing. Ian keeps telling me it will get better. I'm not sure if I can stand it if it gets any better. I guess we will just have to adapt!

Ian walked a friend's ranch with him one day. This friend was in a seven year drought on his ranch. He had pretty well de-stocked the entire ranch because of the drought. After a full day of touring his ranch he asked Ian, "Well, Ian what do you think?"

Ian responded, "You will never grow any grass until you get some cattle."

My friend responded, "There is nothing out here for them to eat!"

What Ian was saying was the cattle bring the grass. You have to do whatever it takes to get the microbes out onto the soil. He went on to say, we all have too much land and not enough cattle. We must have heavy animal density or you don't have any chance of growing more grass.

Another bonus to having your animals bunched up during a drought is the super long rest periods that you build into the system by monitoring the re-growth of previously grazed paddocks. We can go as long as 180 days between each grazing if the need arises just by tightening down our grazing area each day. These long rest periods really shine in a drought.

Another benefit of having all your animals mobbed up during the dry periods is the urine distribution on the small strip of paddock that is presently being grazed. A mature cow may take in 15-25 gallons of water on a hot day. That same amount of moisture is being deposited on your pasture fairly evenly by having them stocked in a tight area. This alone will help jump start your re-growth if you do not allow them to back-graze over the previously grazed area. The livestock are watering your grass with rich nutrient-laden liquid as they graze it off!

Droughts are hard on ranchers. I have read where some ranchers get so depressed they even commit suicide. They may sell the cattle, and quit ranching. What a travesty to give up your passion in life simply because you did not know there was an answer to your drought problem. High Density Planned Grazing is the answer to a drought.

Increased Microbial Action:

If you do not have microbial activity in your soils you're never going to build better topsoil in your pastures. We have noticed that in the areas that are mob grazed, the surface litter breaks down faster versus the areas that are not mob grazed. The plant material is just gone.

The surface litter that is being laid flat on the ground is being incorporated faster by the intense hoof traffic it is subjected to with each strip grazing period. The litter is busted up in smaller pieces, which helps with faster breakdown of the microbial food for the soil bank.

I believe this increased microbial action is the biggest bang that you get for free from High Density Planned Grazing. The hooves turn the soil into gold just by the heavy treading action that the soil is subjected to along with the manure and urine.

We have heard in the past that you may compact your soils by heavy animals grazing too tightly. We have not found this to be the case as long as they are only on the area for a one to half a day period. We are depositing so many bugs from the livestock's rumen onto such a small area, that you can not believe how quickly it wakens up the soil.

One very important thing to keep in mind when thinking about microbes is that they are not active in the winter. So when microbial activity is the highest (the growing season), that is the time you should be concentrating on getting all the microbe food you can onto the soil surface.

Also, you will build topsoil faster in the growing season than you can by feeding hay in the winter. This was a big paradigm change for me. We always focused on feeding purchased hay during the winter days that were unsuitable for grazing. That was our fertilizer program for the coming growing season. This practice does work well, but there is a better way to jump start and improve your forages. Concentrate on feeding those little invisible bugs. That is where your biggest bang for the buck is!

Get down on your hands and knees after the mob has left the paddock for a closer look. What is it telling you? Did you get enough hoof action? How many bare ground spots do you have? Do you see any dung beetle activity?

Drag your foot right through a couple of fresh manure piles. There may already be a few dung beetle dwellers in there. Estimate what percentage of plants were eaten and trampled. Take out your pocket knife and cut a chunk of sod out of the grazed strip and smell it.

How many intact cow pies escaped a cow's hoof being drug through them? If there are numerous intact cowpies, before you move them next time, walk the mob back and forth through the grazed strip a couple of times. This really puts the hurt on virgin intact cow pies. You are getting more of a fertilizer bang with each cow pie by having the cows spread them for you. You don't need a tractor or ATV dragging a harrow. Save your gas money, let the mob do it for free.

The mob is excited about moving. Some will be running and bucking as you take them on their walk through the mine field of manure piles. After several passes back and forth, the field will look like a peanut butter sandwich. We will let the dung beetles have that sandwich!

This same method of walking the cow mob back and forth several times before moving them can also trample a lot of the weeds in areas that were left standing. This is a good thing to work on. Give those areas some animal impact. The severe bruising that the mob gives the paddock brings the soil to life. Neat things start to happen.

I never realized there were billions of little workers in the soil bank working for me. All we have to do is manage our operations to give the soil critters a good solid meal every day that we possibly can.

I will say this again, stop using nitrogen fertilizer if you want a high population of working active microbes.

I believe Ian is right, we should be calling ourselves Microbe Farmers instead of Grass Farmers! Concentrate on growing a healthy community of microbes and you will grow more grass than you know what to do with.

Improved Water Holding Capability:

We capture all moisture and hold it right where it falls with the rested fully re-grown swards of grass. It compares to dumping a bucket of water on a bale of dry straw. It just sucks it up and stores it. The winter rains are the only time we get runoff on our farms. This is because the plants are dormant and the soil surface may be frozen, which keeps the rain from soaking in.

The only downside to High Density Grazing that I have found is that you do not get much runoff into your ponds during the growing season because of the huge sward of grass that captures the rain and holds it in place. The ponds may get low in a severe prolonged drought. So far in our operation, the ponds have filled back up in the winter period.

The long grass leaves have more surface area to capture dew each morning. This alone can help keep grass green and vegetative in a droughty time.

The fully re-grown roots under the plants add tons of humus when the top is removed. This humus is the life of your pasture. It has many times its weight in water holding capability. All the mature roots that die off with each grazing are nice deep tunnels for soaking up water.

We have read that after years of High Density Grazing you may start seeing springs and creeks flowing again due to the increased holding of water in place on your pastures. I saw my first dribble coming out of a hillside at the bottom of an old ditch our second year. I could not believe there was water oozing out of the ground simply because it had not rained for three weeks.

This ditch area sits a quarter of the way down on the side of a hill. There should not have been any moisture trickling out of that ditch with no rain. This trickle continued for several weeks then stopped. Maybe as we continue improving our soils we will see more of this in the future.

Dung Beetles:

Dung beetles, also known as tumblebugs or rollers, are a tremendous asset to your grazing operation. There are thousands of different species of dung beetles worldwide. The United States has many different species of them. Most species fall under "the tunnellers" group.

Another group, "the rollers," bury a dung ball for either food or making a brood. One male and one female will be seen around the dung ball during the rolling process. The male will usually do the rolling, and sometimes the female may help. They make a nest at the end of the tunnel for their dung ball. They will then mate underground.

The female lays her eggs inside the dung ball, which is the food source for the larvae. About 50% of the dung ball is used up by the larvae during the process of transforming into an adult. I have seen dung balls that were actually bigger than the beetle rolling it.

A third dung beetle group is called dwellers. They live in the manure. Previously, I never saw much dung beetle activity on any of our farms. We now have one species that actually builds a small castle with mined dirt chunks. It comes up out of the ground when there is surface moisture. It resembles a crawdad hole. The more species you have is a good thing! We have seen three different sizes of dung beetles so far on our farms.

There is one small beetle about the size of a housefly that is always tunneling around in the manure. I'm guessing he is a dweller. We have another one that digs a hole about the diameter of a pencil in the soil surface. There may be holes that size scattered all over in the vicinity of the manure. We have a larger one about the size of a June bug that excavates dirt to the surface. The excavation usually takes place right at the edge of the manure pile. Sometimes I have seen fresh dirt pushed up right in the center of the manure pile. This large one is very shiny green colored.

Some dung beetles will try to steal dung balls from other dung beetles. So they move the ball rapidly away in a straight line from the manure pile to keep it from getting stolen. They have a strong sense of smell, which guides them to the manure. The adult dung beetles live off the liquids from the manure.

I saw a glass sided, soil filled window that was used to view dung beetles. You could see the many tunnels that the beetles had made and deposited dung balls in. It was amazing, the tunnels went everywhere and to various depths. Think of your pastures as this glass window. An active population of dung beetles could really aerate your soils.

Dung beetles increase soil fertility by burying dung deep in the soil profile. They are getting the manure down into the plant root zone. Instead of the roots getting the nutrients fed down from ground level. You are also getting tunnels dug by the burying of the dung ball and from the new dung beetle emerging from his nest.

By having the dung beetles incorporating the manure into the soil, there is less surface fouling of your pasture. The quick disappearance of the manure pile allows that spot to be grazed again the next grazing. Add up all the old manure piles over your pastures and we are talking about a significant area of your pastures not being utilized.

If the manure pile is left on the surface, studies have shown that up to 80% of the manure nitrogen is lost through volatilization. Eighty percent of your nitrogen is gone because you do not have an active population of dung beetles! If that doesn't get your attention about the importance of dung beetles, check your pulse, you're not breathing.

Flies are attracted to manure piles and lay their eggs there. More flies hatch and the cycle keeps repeating itself all summer. What if there wasn't a manure pile for the flies to lay an egg on? Good-bye flies right? I feel sorry for livestock during July, August and September when they have to swat at the ornery flies that love preying on them. The constant fly irritation costs the livestock producer in many ways. Livestock lose weight, and other health issues will start to affect them as well.

The dung beetles feed off the liquid part of the manure pile, which reduces the water content of the pile. This effectively ends the incubation of larva that will eventually end up as parasite loads in your livestock. So even if they don't bury all the manure, just removing the water content kills the larvae. Also, as dung beetles process a manure pile they actually compete with the fly larvae for food, and damage the fly larvae as well.

Dung beetles are a free species that are working to make your operation more sustainable. They can not tolerate Ivermectin wormer. It kills them. By keeping all chemicals off your operation you will have a great place for dung beetles to call home. I have noticed the more intensely we graze, the more dung beetle activity we see.

The other day we were walking one of our farms that

had been mob grazed. We were checking on plant re-growth and started noticing black critters darting for cover as we walked. My wife thought from a quick glance that they were big black crickets. They were dung beetles working their little hearts out. What really surprised us was the awesome dung beetle activity still going on 90 days after the cattle were re-moved. There were drilled holes in the ground everywhere with black beetles sliding down them as you approached their location. The soil surface looked like you took a cordless drill and went crazy punching holes with it. They were even scam-pering down in the cracks in the ground from the drought we were experiencing. I wonder if they were storing dung balls down there as well?

The Mob and the Bird Factor:

Our place has gone to the birds! The cow bird flocks we are seeing are breathtaking. Never have we seen large flocks of birds exploding up out of the mob like we do now. These birds are picking, scratching, consuming fly larvae and other insects that are attracted to the cattle. I have never seen birds so aggres-sive around cattle. They are scratching right next to the cattle as they graze. They land on the cow's back for convenient perches.

We have more Meadowlarks than ever before. Many other species of birds are in the area as well. The wild turkeys are spreading out the cow manure looking for bugs. They had better be fast to catch a dung beetle! The increased bird activity is another missing link that has been connected on our opera-tion due to High Density Grazing.

I've noticed that when the mob is moved onto the fresh strip of rested grass that the bird flock will explode into the air and follow them onto the fresh paddock. The movement of the cattle grazing the fresh paddock stirs up the insects in the grass sward. The birds go on a feeding frenzy picking insects out of the air. It's pretty neat stuff to watch.

The birds also will scratch through the duff on the

grazed paddock, aerating it as they feed. What did it cost me to get that flock of valuable birds? A management change of moving the wire and keeping the cattle mobbed up.

We are convinced that if you truly want a profitable grazing operation, there simply is not a more sustainable method of achieving your goal than with planned High Density Grazing. In a year and a half with this type of grazing we have seen more positive changes on our farms than in the previous 20 years combined!

The cost? Nothing but a change in management styles, planning, and some purchased temporary posts and reels.

The exciting part of this whole concept is that we don't know how far we can go with it in regard to stocking rate, increased quality of life, profit, wildlife, plant diversity, better soils, better mineral cycling, effective water storage, and stronger neighborhoods.

Chapter 27
Sell Your Mower, Mob Those Weeds!

Weeds are a common occurrence on most farms during some part of the growing season. This is just a fact of nature. There will always be some weeds that take up residence in a particular spot on your pasture.

Weeds have been called, "Nature's Band-Aid." The weeds usually come up where there has been some bare dirt exposed. They waste no time in filling in these bare areas. They are opportunists. Grazing management can have a huge effect on weed populations.

Excessive weed populations are a red flag telling you that something is wrong with the management of the land. I used to think that the only way to combat weeds was to mow them off, and over time you would get rid of them. This was not the case though. There are millions of weed seeds in the soil bank, and until you change your grazing management you will always battle them.

Most people around us mow their weeds in the July-August time period. But, they are also mowing off tons of grass and legumes along with the weeds. The pastures dry up, and they are stuck with no grass and a pasture full of weed stems. The weeds just laugh and are back the following year.

If you mow in this late life cycle of the weed, you are also spreading mature seed into areas that may not have any weeds. You are propagating them. For heaven sakes, stop!

Driving by a mowed field may look pretty if you like a mowed landscape, but a closer investigation will show the true content of the pasture - weed stems. The grass that got clipped when you mowed the weeds is now having to re-grow, assuming there is enough moisture available for the grass plant to do so. Guess which root structure is stronger?

The weeds will be back. The grass roots are even further weakened and will have an even harder time competing with the weeds when growing back.

For the last several years we have been managing weed problems with high density mob grazing. With High Density Grazing the weeds really get run through the mill. They are eaten and stomped on to the ground along with the grass.

Breaking and stomping weeds is a huge source of valuable organic matter that builds the soil when it breaks down. With mowing weeds you are not getting any animal impact on the soil surface to facilitate the breakdown of the weed material into the soil bank. Ironweed, curly dock, common and giant ragweed, milkweed, mares tail, Canadian thistle, burdock, Queen Anne's lace, goldenrod, horseweed, common mullein, purple loosestrife, velvetleaf, field bindweed, along with many others are being consumed or stomped.

Our grazing density has ranged from 100,000 to 500,000 pounds live weight/ acre. That is equal to100 to 500 cows per acre. The heavier the grazing density, the better results you will have if you want them all eaten or stomped.

One of the abandoned idle farms we leased had several fields that were almost solid goldenrod. With previous lighter grazing densities the cattle never touched one plant. The cattle grazed what little grass was between the plants, and the goldenrod was completely untouched. Each year the goldenrod got thicker, and the grass stand thinner..

For the last two years this field was grazed using High Density Grazing. It was amazing what the cattle did to the goldenrod. We had the density high enough that they ate and stomped the majority of the weeds flat on the ground. After two years of this practice other grasses and legumes are starting to appear in the field, and the weeds were getting fewer. It's the same story with ironweed, except the cattle ate it better. They even ate a lot of the ironweed stalks, where before, with lighter grazing densities, when we removed them from the field it looked like an ironweed field that had not been touched.

We are starting to notice that with longer rest periods between each grazing, the weeds have a harder time competing with a strong mature sward of grass. The huge grass sward shields weed seedlings from the sun, which prevents them from competing with the grass. A strong grass sward can hold its own against any invasive species.

We have numerous pastures that previously held large amounts of weeds that have been replaced with strong swards of grass. Our best pastures in drought conditions are the ones that have the longest rest periods between each grazing period.

The farms that are only grazed twice per growing season have phenomenal stands of grass with little weed presence. I know a lot of you may be thinking that long rest periods will have a negative effect on the quality of the grass sward. I thought the same thing initially, but have found that in our region (38" precipitation in a normal year) the quality stays pretty high as long as we used High Density Grazing.

My wife and I walked a farm one night that had had 90 days of rest (page 201 compared to the same field page 200). The quality was just eye popping. There was lush new grass in every paddock that had very little seed heads. The diversity was outstanding throughout the farm. With one year of two grazing periods using a high density of 4-500,000 lbs per acre (400-500 cows), we changed the whole farm around.

This spring this farm was primarily rank fescue, weeds and broomsedge. Walking the farm we spotted orchardgrass, timothy, redtop, numerous varieties of summer annuals, lespedeza, red clover, white clover, crabgrass, barnyard grass, big bluestem and numerous other grasses. Just looking at the awesome pasture made us want to grab a bowl and fix a salad! All of these wonderful results were made with no added inputs of any kind, just a change in grazing management.

The only noticeable broomsedge left on this farm is directly under the permanent wire divisions of each paddock. These were areas that were protected from animal impact and manure distribution.

Another amazing result we saw was the small dirt spots exposed from the intense animal impact had no cracks in the ground surface even after 80 days with no rain and 100 degree F days.

We have other farms that are not set-up to mob graze yet, and these farms have cracks on the surface even where there is grass covering the soil. What is keeping the ground from cracking open? Could it be a mixture of heavy animal impact, manure and urine, stronger more dense root systems, dung beetles, turbo-charged microbial activity? Maybe all of these things combined are having a positive effect on the ground resisting cracking open. I do remember the smell when you approached the fresh grazed strips. It was a very earthy, wholesome smell, not like a feedlot at all.

It is a nice feeling now to look out at a field of weeds and smile. We now have a management system that runs on solar power, not fossil fuel, which turns weeds into forage. Weeds have long tap roots that can mine minerals from very deep depths in the soil, which are inaccessible to the grass plants. What we have noticed with most weeds that are completely bitten off by the cows is that when they re-grow during the rest period there will be four to eight plants coming up where there was previously only one stalk, more forage! These tender new plants are immediately consumed when exposed to a mob of cows.

Let's quit calling them weeds. Instead we will call them forbs. By adding forbs to the cow's diet, you are also adding more diversity of minerals into her body, which is a good thing.

Go out and string up some portable grazing strips. Monitor the results you get and adjust your grazing density to harvest all the energy you can from these forbs.

Observe the ground surface. Look at what's going on

around you. Look at the graze-to-trample ratio. Watch the cows graze when you turn them into a new strip. They should walk in, put their heads down, and start eating. There should be no walking around or bawling, just the sound of grass being snipped off. This is fun and exciting stuff folks. You don't know what you are missing!

My brother-in-law from San Francisco was accompanying me while we moved the mob one day, and he commented how sweet the air smelled as we walked over the freshly grazed area. I gave this mob to him and my sister for three days to move from one strip to the next. They both fell in love with the mob movement and observing the results from each grazing. These were folks who had never been around cattle before, but were very aware of what was happening in the grazed surroundings. If they ever wanted to give up the city life they would make excellent graziers!

I have to end this chapter with a story about my brother-in-law. He was helping me move a mob one night in the dark. I had my coon light strapped on and had offered him one as well. He turned down the offer of wearing a light, and said that he could keep up fine by following me. He was a little bit reluctant to be out in the middle of the mob, in the dark.

The temporary paddock gate was at the bottom of a large hill that had several trees growing along the side of it. I knew what was going to happen when we started down the hill with 250 cows following us. I immediately found a large tree and got behind it. When I shut my coon light off, that really got my brother-in-law's attention. He alarmingly asked, "What did you do that for?"

I replied, "Well, you're getting ready to have 250 missiles that weigh 1200 lbs each come charging down that hill, and a bright light blinds them. I want to make sure they see this tree! If I were you, I would get behind this tree with me."

No sooner than he had gotten behind the tree when we had 1200 lb cows shooting by us on both sides, running, bucking, excited to move to the next paddock. It was quite a sight in the moonlight to watch the mob charge by us. I think my brother-in-law was holding his breath.

As we joined the cows at the bottom of the hill, we walked all through the mob checking out the calves. The cows were literally just standing around, and you had to almost push on one to move so that you could move forward in the mob. My brother-in-law was amazed how gentle the cattle were with us right in the middle of them, and commented later that the closest he had ever been to cattle was looking at them from across a fence.

He really got a charge out of the whole night moving mob episode.

I bet next time he will be looking for a tree of his own!

Chapter 28
Gamagrass Loves Mob Grazing

Eastern gamagrass has been described as the ultimate summer "ice cream" grass.

It is a native warm-season grass and a genetic first cousin to corn. Apparently, the early settlers knew that wherever they saw a stand of eastern gamagrass growing the soil was rich enough to grow corn.

Anibal Pordomingo says it is the only warm-season perennial that you can finish cattle on in the hot mid-summer.

The praise for this grass has been loud and long. It appears that the only people with something nasty to say about this alleged wonder grass are those of us who have tried to grow it using the conventional advice.

We planted a field to gamagrass in 1994 by conventional tillage. The pre-soaked seeds were drilled in the ground with a corn planter. I followed the advice that was given and deferred grazing the first year.

The second year I was told to put down 60 lbs of nitrogen right at green-up, and I followed this advice. What I grew was rank fescue and weeds.

Each subsequent year the fescue got stronger, the gamagrass weaker. It seemed the more we grazed it, the more the fescue had a strangle hold on it. The livestock would only pick through the canopy, barely touching the rank forage below.

I burned it a couple of years, which helped some, but I hated seeing all that organic matter evaporate into thin air. Continued burning causes plant spacing to increase not decrease. Bare ground is a bad thing in a solar collecting business like grass farming.

I had a fellow tell me that I made the serious mistake of grazing it! He said that you should only hay it to maintain a good stand. That was not an option for us, since we do not put hay up on any of our ground.

I was also told gamagrass required lots of nitrogen to maintain a healthy stand. This left us out. We now refuse to buy commercial nitrogen because it kills the active microbes in our soil and leaves the pocketbook empty.

The last couple of years were so bad that you could hardly see the gamagrass anymore. We finally decided this high input, low output gamagrass field had to go. As fragile as this grass was supposed to be, we decided to kill it by grazing it at the wrong time of year.

In the fall of 2006, we decided to try and mob graze it with dry cows at a stocking density of 500 cows per acre. We purposely strip grazed it off in October because this is supposedly a bad thing to do to a warm-season grass. They store all the energy they need to survive until the next growing season in the bottom eight inches of the plant, and we were always cautioned to forego fall grazing and to always leave a high residual.

No doubt, 500 cows to the acre grazing this grass to one inch in height would be the death of this worthless grass!

The dry cows were strip grazed across the bottom with a back fence kept in place daily to maximize animal density. The cows consumed everything!

They busted up the old gamagrass crowns with their hooves, and the whole bottom looked like a field that had been hit with manure pies. Literally every two feet had a large manure pile. We thought to ourselves, "Boy, we have killed it now!"

But those manure piles soon faded away. By the time April 2007, rolled around, the whole bottom had a golf course green color to it. Much to our surprise, there were small gamagrass plants coming up everywhere along with a thick-ened, invigorated stand of fescue.

We were dismayed. We had failed and so redoubled our efforts.

On May 8, 2007, we turned in 140 dry cows to mob graze it again. The cattle ate everything and stomped what they didn't eat. The manure distribution was awesome. The whole bottom looked like a sheet cake of manure.

We thought to ourselves that surely we had completed the job of finishing off that weak stand of fragile gamagrass.

However, we could hardly believe our eyes as the weeks rolled by. As the weather warmed up, the gamagrass kicked into high gear and completely took over the whole field, domi-nating the fescue. The original gamagrass rows were com-pletely visible again. Our old weak stand now looked like it had just been planted a year earlier. It was completely rejuvenated. The gamagrass leaves were very dark green, and looked like you had put 100 pounds of artificial nitrogen on the field. There was a carpet of lush lespedeza filling in the rows between the gamagrass. It looked like a salad bar of green vegetables. This pasture was at last the beef finishing quality we had heard about!

How could we do everything so wrong and have every-thing come out so right? We do not have all the answers, but we do have some possible theories.

1. The high density mob grazing apparently stimulated the old gamagrass stand back to life by creating the conditions the grass had evolved under in the buffalo herd days. This is exactly what happened to the grass for centuries with the massive buffalo herds.
2. The May mob grazing hammered the fescue and set it back enough to release the gamagrass for its coming summer peak growing period.
3. The excessive manure distribution gave the gamagrass all the nutrients that it needed to fully express its growth.

4. There was not any dead thatch to retard growth or sunlight.
5. The ground was basically bare after the May grazing, which collected more heat and stimulated the warm-season gamagrass to come up even faster.

Here are my conclusions from this experience:

If you want to revitalize an old pasture, mob graze it with dry cows. Do not use stockers, finishing animals or lactating cows as they will not be able to meet their nutrient requirements.

No purchased nitrogen is required to grow an extremely healthy stand of gamagrass, even though it is a nitrogen lover. Who fertilized it two hundred years ago? Animals, not people.

With nitrogen prices soaring to 57 cents per pound in the spring of 2007, the high density manure cake that covered the field was very economical.

In just one year we turned an extremely rank field into a dynamic, healthy, diverse grass/legume finishing pasture with no purchased inputs. The only tool required was High Density Grazing and an aggressive attitude trying to kill it! By breaking the rules, we finally got the grass to work as it had been advertised.

We love our gamagrass field now that we have learned how to manage it using a sustainable method such as the MOB.

Chapter 29
Calving with High Density Planned Grazing

When we first started Holistic Planned Grazing, I had my doubts about calving in a mob. I envisioned calves being abandoned by their mothers after being all mobbed together. This has not been the case.

We like calving in a mob for several reasons. First of all the cattle are tame by being moved twice per day, which makes them easier to handle if you have a problem. When we have to sort out a cow that is having a problem calving, it is very easy to do. We walk out with a polywire reel, and we extend the polywire between us and walk the problem cow and a buddy cow over to the edge of the current temporary fence. Next we walk the pair up to the area of the strip where there is a temporary gate. We immediately erect a pen using the same polywire reel that we used for walking them to the gate by stepping in some temporary tread-in posts.

We make this pen about 20' x 20' and energize it by draping the reel over the existing hot wire fence.

On the opposite side from the gate we have a temporary polywire lane going back to the corral. The cow stays fairly calm as long as she has a buddy with her.

We then walk the pair to the corral and sort out the cow that is having a problem. The buddy cow is then walked back to the mob after we are done working the problem cow.

One good result from calving in a mob is that the new calves are very easy to find and ear tag. If a cow has calved in a large pasture and she has hidden the calf, you may not find it for several days. Have you ever tried catching a calf that is two to three days old? They are as fast as greased lightning.

We will always tag the new calf first before moving the mob. Once all new calves are tagged we move the mob to their fresh paddock strip. The new mothers will almost always stay behind with their calves. We walk the new mother and calf over to the edge of the polywire fence, which is not energized because we left it unhooked when we moved the mob. I reach over, pick up the polywire fence and lay it over the cow's back, which puts her in the new paddock with the rest of the mob.

This little trick keeps you from having to walk the cow all the way up to the gate, which she is reluctant to do because of her new calf. Most of the time the new mother is so focused on me that she does not even see the wire come over her back. It sure makes moving new mothers and their calves easy.

Another advantage of calving in a mob is that the predator issue is no longer a concern. We have not had one predator attack with our mob of cows. The cows are tightly bunched up, which does not allow a predator much of an opportunity to slip in undetected.

Our guard dog is not even permitted in the mob by the cattle when they are calving. They run her out. The massive buffalo herds stayed in mobs for predator protection. The weak animals that could not keep up fell prey to the predators. If you look out across most pastures today the cattle are all spread out. The calves may even be lying by themselves unprotected. Nature's system works the best. The cattle are mobbed up for protection, the grass gets maximum animal impact, and calves are protected.

We have had two cows that prolapsed over the last two years. They both died leaving us with bottle calves. Rather than sell the calves for $100 as newborns, we bottle fed them for three weeks and turned them back in with the mob. The calves both did fine by robbing milk off of other cows and grew out nicely. The orphan calves always snuck in from the back and stole their dinner.

Once the bottle calves were three weeks old, they were very aggressive for searching out their milk. They just kept moving from one cow to the next until they found one that would let them suck. They had multiple mothers!

By having all the cows in a mob, the orphans would go undetected most of the time when searching out dinner. With traditional bottle calves you can never get the growth on one that a cow will give them. The milk from the bottle just is not a very good replacement for good high quality cow's milk. The calf being raised by a cow will get multiple feedings a day. With a bottle it becomes a real chore to feed them more than twice per day.

While we are talking about calving, I want to cover how we tag calves. The cow should tolerate you putting an ear tag in her calf. But, there are always horror stories of people who get killed every year by cows mauling folks to death when attempting to catch the calf.

When I approach a new calf I am always on my 4-wheeler. I want something that I can dodge behind, or get on, to offer myself some protection.

Any cow that is so protective that she charges you should be sold. This can not be tolerated for any reason. She is bigger and stronger than you, and she will always win the fight. You simply can not afford to have this type of cow on your farm. She will put you in the hospital or worse. Now you're out of business with hospital bills, missed work, etc. I have heard people say, "Yeah, she is pretty snotty when you get around her calf, but boy she sure throws a good calf." I don't care how good of a calf she has, it is not worth the risk of getting badly hurt.

We ran some custom grazed bred cows five years ago from out west. These were range cows that had never seen a person on foot, only from horseback. The day they arrived I knew we were in trouble. They ran off the truck and disappeared over the hill in a cloud of dust.

I decided to go introduce myself to them, and gently

walked over the hill. I never got any closer than 200 yards. They took off back across the pasture in a stampede.

Two of them jumped the fence and took off down the road. Well, after about a month they would tolerate my driving by them on a 4-wheeler without stampeding. They started calving, and the owner wanted every calf tagged. The only way you could tag the newborn calf was to ride by it at a very brisk pace and reach over and grab the calf without stopping. The cow would literally try to kill you if you got off the 4-wheeler. As soon as I got a good hold on the new calf, I would throw the calf in my lap and give the 4-wheeler the gas with the cow right on my behind. I would drive as fast as I could several hundred yards away, tag the calf, then briskly cruise back by the new mother. As I passed her, I would slip the calf off on the ground and drive off.

I used this process with all 120 of those cows. They would flat eat your lunch if they got a chance. One of those cows came down with foot rot. The only way we got her in the corral was to use myself as bait. She chased me in the corral, and my helper closed the gate behind her. You could not get in the pen with the cow without her putting you over the corral.

I called the cattle owner and told him that this cow had to leave the farm. He asked, "What's wrong with her?"

I replied, "Come out in the morning and see for yourself."

When he drove up to the corral the next morning he parked his truck right by the corral.

As soon as he got out of his truck, the cow just smashed into the corral panel he was looking through. It was slobbering, bellowing, and pawing at the dirt.

He replied, "She is kind of a hot one, ain't she? Let's load her."

I replied, "I will open the trailer gate. She knows you. Why don't you run her in?"

It didn't work exactly like that. Luckily, I have a very good corral system, which allows you to advance the cow up the ally by keeping six-foot corral gates between you and the cow. We finally inched her onto the trailer by keeping gates between us. I was never so glad to see a cow leave our place.

We stopped calving custom cows after this. What I found out was that some folks' idea of a tame cow and my idea of a tame cow did not match up very well. Some people will tolerate any temperament to get a calf on the ground. Life is too short to subject yourself to this kind of misery.

Why do we tag calves? By having some kind of ID on the calf you can match them back to the cow and the bull. When you have a cow that loses a calf for whatever reason, that cow is marketed off the farm right away. Every cow must raise a calf every year unassisted in her environment without any added inputs other than mineral or she does not have a home on our farm. If you do not stick to this selection process religiously you will end up with a whole herd of high maintenance cattle that will break you. What if a calf gets out and enters your neighbor's herd? A one dollar ear tag would be pretty cheap insurance in reclaiming your calf.

When you sort your calves at weaning, it allows you to match the performance of the calf back to its mother. If you have a dink calf at weaning, well, that cow is a good candidate for culling. She will probably give you a dink the next year.

A poor performing cow eats as much grass as a good cow. Why not have the option of selecting for the good one? Ear tags allow you to keep a very accurate count on the live calves. Instead of just counting heads and hoping they're all there, if one is missing, you know exactly which one it is.

If you have to doctor a calf, how are you going to sort out the mother if you have to pen them up for awhile if nothing has a tag on it? When you keep replacement heifers, they already have a tag in them, which gives them a cow herd number to trace their performance.

Chapter 30
What Will the Neighbors Say?

If you don't have your neighbors talking about you, you're not pushing the boundaries of your operation enough. Never be satisfied with achieving an average performance. There is always a better, more profitable way of doing something.

We were in a drought in the summer of 2006, when a neighboring cattle fellow stopped by to investigate what I was doing. This particular farm stretched along the blacktop for about half a mile and got a lot of stares from onlookers.

This fellow walked up to the fence, and asked me what I was doing. I had been strip grazing a dry cow mob with twice daily moves all during the drought. He commented, "The rumor around the coffee shop is that you are out of grass, that is why you're penning your cattle in those little bitty strips."

His next comment was, "Why don't you give them the whole dang farm? The cattle will do better, the farm will do better, everything will do better." His voice was getting a little louder toward the end of his sentence.

The guy was about half angry that I was strip grazing those dry cows. I patiently explained to him what I was doing and the positive benefits that I was seeing from doing it for such a short time period.

He scoffed at my explanation of the process and replied, "Too dang much work." He walked back to his flat bed truck and drove off with a 1800 lb big bale hanging off the back of his truck. He had been feeding hay for the whole month of July

213

and was madder than a hornet that I was still grazing and had worlds of grass left.

Which is more work? Taking 10 minutes to move a polywire or feeding a big round bale every day?

I know which is more profitable and it sure isn't feeding hay in July.

On another farm in 2007, a neighbor stopped and stared at the dry cow mob that consisted of 140, 1400 pound cows on an eighth of an acre. They were grazing right by the road. I had them stocked at 1,568,000 lbs per acre. This was about 10:30 a.m. when he came by. He got out of his truck and walked up to me with a look of concern on his face. He commented, "Greg, what you're doing there ain't right. You're stressing those cattle by having them grouped up like that."

I looked out at the mob, and every cow was standing completely still, grazing off huge mouthfuls of fresh grass. Not one cow was bawling or paying any attention to us. All you could hear was chomps and grass being ripped off, a very neat sound. Every cow had a nice slick shiny hair coat and was in excellent body condition. I turned around to him and replied, "Yep, they look pretty stressed to me."

I had already moved them twice that morning to fresh rested strips. The neighbor assumed they would be there all day I guess, but they were moved five times that day to fresh strips.

This fellow owned around 1000 acres of pasture and grazed around 60 cows on it. This was June and his pastures looked like a parking lot. The hot dry summer had not even started yet, and his cattle were looking at a long summer ahead of them. Those 60 cows had access to the whole 1000 acres all the time, with no rest for the plants to recover from grazing.

What is amazing to me is that people can physically see what is happening right in front of them with this mob grazing, and they still don't get it. I don't think it is ignorance. It is either pride, a closed mind, resisting change, jealousy, being scared or laziness. It may be an assortment of all these combined.

The old saying, "The proof is in the pudding," sure describes what you see when you walk out into the mob and then examine where they were previously.

Chapter 31
The Landowner Mob Move

In the summer of 2007, I invited a prospective land-owner over to show him our High Density Grazing system. He owns a 145 acre farm that is sitting idle, and wanted me to set up a grazing system on it. I figured the best way to explain to him the high density system was to actually go move the mob.

This fellow is not a cattle person. He lives in St. Louis and uses his farm for recreation and to hunt on. Once in the pasture, we were walking over strips that had been grazed for about a week. There was a beautiful carpet of grass, legumes and forbs coming up, and there were old broken up sheet cakes of manure everywhere.

When we reached the mob, the area they were on was covered with fresh manure. The landowner with an astonished look on his face commented, "You mean to tell me that the area they are on right now looked like that un-grazed piece in front of them 12 hours ago?" He simply could not believe that this mob could turn a piece of pasture into a manure carpet in 12 hours.

He commented, "I could not have done a better job of mowing this strip with my lawnmower and it doesn't fertilize while it's mowing!"

I commented to the landowner, "Would you like to move the mob?"

He was almost bouncing with excitement and asked, "Would it be all right?"

I gave him the instructions about the correct procedure

for unhooking the reel without getting shocked. I told him when you remove that reel and start to pull back the temporary polywire, not to dilly dally around. I advised him to roll the wire back 100 feet as quickly as he could and get out of the way.

I could tell the landowner was a little nervous when all 140 cows were anxiously walking right behind him over to where the reel was hooked onto the fence. He did exactly like I told him. He trotted with the reel out into the fresh paddock giving the mob a nice opening to enter the fresh strip. It was so neat watching him move the mob. He had the biggest grin on his face.

The mob charged through the opening, then immediately they all stopped, and heads went down grazing.

I walked over to him after the last cow had walked through the opening. He was almost giddy with excitement. He commented, "My gosh you could feel the heat of the mob as they passed me, and the power expressed by the mob was simply amazing."

We went back to the house to have a cold drink and visit. He could not stop talking about what he had just seen. Here was a fellow who had never been exposed to any livestock, yet he completely understood the process of mob grazing along with its many benefits.

Before the landowner moved the mob, he was kind of lukewarm about me setting up a grazing system on his farm. His attitude was basically, what will you pay me to lease my idle farm? After I explained to him what it was going to take to put in all the necessary items to set up a grazing system, he had a change of attitude.

He basically offered us a free seven year lease on the farm. His farm had no water and no fence, only 50 acres was open grass pasture, the rest was brush and woods. The farm is located 27 miles from my driveway, way too far for me to operate a grazing system on a daily schedule.

I had two young men in their early 20s who lived fairly

close to the farm that were eager to help me develop it. They had read my first book *No Risk Ranching* and called to see if I would let them work for me for free to learn some of the tricks of the trade.

These young men worked for me and learned the basics of mob grazing along with many other things. If I got the lease we were going to be partners in the operation. I was going to supply the equity to put in the inputs, they would supply the labor and daily moves of livestock.This would be a great way for them to get their feet wet with leasing land and grazing cattle with no equity or risk, just a lot of sweat.

What scared me a little about the land lease was the landowner was 78 years old. He even commented that he might not live seven more years, which was the length of the proposed farm lease. But he immediately came up with a solution. He offered to put in the written lease that if he died, the lease would carry on until the end of the contract even if his kids sold the farm. That's a pretty strong commitment from a landowner! He really wanted to see the chips fly with a grazing system on his idle farm.

Chapter 32
Combining All the Herds

In July 2007, Ian Mitchell-Innes sent me an email asking if we would be interested in putting on a grazing seminar with him on our farm. I about fell out my chair when I read the email. Here was the leading world expert in Holistic, High Density, Planned Grazing asking if we would host a grazing seminar with him. I immediately forwarded the email to my wife. Jan fired back an email, asking me if this was some kind of hoax. She could not believe it either.

I sent Ian an email telling him we would be honored to host a grazing seminar on our farms with him. Ian and his wife were coming to the United States in October, so we set the date of the seminar for the last Saturday of that month. Ian and I decided to have four hours of classroom presentations describing Holistic, High Density, Planned Grazing followed by lunch. After lunch we planned on having three separate pasture walks on our farms showing three separate High Density Grazing systems.

We placed several ads in *The Stockman Grass Farmer* and Jan started putting together all the arrangements for the meal, coffee and doughnuts. We rented the local Lions Hall because it had a large seating capacity.

Ian and his lovely wife, Pam, arrived at our house to spend several days before the seminar. We spent a day with Ian giving him tours of our farms. As we visited the various farms, Ian gave us his thoughts on what he saw at each one.

I had never been around someone so knowledgeable

about managing a grazing system in sync with nature. In South Africa, Ian has an advantage over us in the United States in that they still have the large herds of wild animals that live in the wild. Ian gets to study these wild herds and see how they exist in nature without mankind's intervention.

Ian is constantly studying the actions of animals and people. He commented that you can learn a lot about someone just by studying their body motions and actions. I learned many invaluable things from Ian during the days we spent together on our farms.

October 27, 2007, the small town of Harrisburg, Missouri, was invaded by over 200 people from 19 states who had come to learn more about High Stock Density Grazing on our three farms. Nearly every parking space in town was filled with cars with out of state plates. Needless to say, the locals wondered what the big buzz was about that had taken over their town.

The Lions Hall was packed. Additional chairs and tables had to be set up to accommodate the walk-ins.

Ian led off the seminar with a dynamic talk on the social, financial, and environmental benefits of Holistic Planned Grazing.

I explained the High Density Grazing results that we had encountered in the two years since changing over to the new management system.

Ian followed up with an additional presentation of the basics of setting up a High Density Grazing system.

After lunch was served the huge caravan of vehicles proceeded to the Judy Farms for an afternoon of pasture walks.

The first stop was at a leased farm that had been mob grazed the previous year and was being used that year to graze stockers. Graze-to-trample ratios were explained along with a paddock move showing the simplicity of daily moves with a portable water tank, reels and tread-in posts.

Next the whole group walked up a gravel road an eighth of a mile to the hair sheep farm. I explained the management of the hair sheep and the development of the parasite resistant flock. Lots of excellent questions were asked concerning the economics, management, forage, wintering, predator protection, fencing issues, marketing, and of course, the hair sheep.

Next the caravan proceeded one mile down the road to the farm that was stocked at 368,000 lbs per acre with cow/calf pairs. (That's 70 cow/calf pairs.) The group witnessed the mob being moved onto a fresh strip of grass.

Then we examined the graze-to-trample ratio of the grazed paddock, manure distribution, plant species, tire-tank watering system, and many other topics. Again many great questions came from the crowd. Ian commented that he had never been on a pasture walk that showed so much enthusiasm from that large group of people.

The final stop of the day was on the Judy farm where we were grazing bred heifers and Tamworth pigs. The Tamworths are a truly remarkable hog that can fatten on grass and clover pastures alone. The Judy farm was grazed the whole growing season with custom grazed bred cows at a density of 500,000 lbs/acre (that's 150 animals).

(Here's how that was calculated: We took 150 animals x 1400 lb cows = 210,000 lbs. The half day strip grazed paddocks were 4/10ths of an acre. That is 210,000 lbs per 4/10ths/acre. If we double that number it comes to 420,000 lbs per 8/10ths of an acre. So, to get the full stocking density per acre, not 8/10th an acre, we have to divide 420,000 by 10 to get our missing 2/10ths number to complete the calculation of an acre. It takes 10 tenths to complete an acre.)

The pastures had fully grown back with only 1.5" of rain since June 28th. The energy that the mob injects into the soils is unbelievable. Massive amounts of forage are replenished with very little moisture and no purchased fertilizer. The diversity of the grasses on the Judy Farm have exploded since being mob grazed. Pasture pugging issues, watering systems, rest periods, and custom grazing were explained in the control of using the heavy stock densities.

Ian had walked all of our farms with me before the seminar. He gave me lots of invaluable tips to help us along on our High Density Grazing journey.

That evening I asked Ian what he thought about our operation as a whole and he made a comment to me that floored me.

He said what we were doing was not sustainable!

It was like somebody dropped a brick on my head.

Ian proceeded to explain his statement. He said that with three separate grazing systems being managed at once, that we would eventually burn ourselves out with the constant workload required by them.

Man did I need to hear that!

We had a million reasons why we were running three herds.

Some of the reasons included:

1. The distance between the farms, which I had already envisioned as a distance too far to walk them.
2. Stockers running as one herd for better gain.
3. Yearling heifers that we did not want bred.
4. Bred heifers were being grazed as a herd to remove some pressure during the drought from the large farm where the big cow herd was.

What we were effectively doing was grazing in three spots at the same time. The results were shorter rest periods, less animal impact and three times the work load.

These farms are three to five miles apart and the hauling bill we envisioned getting the large herd to the next farm gave us nightmares.

Ian calmly explained that cows have legs that are built for walking. Let them walk. Why go through the expense of hauling them?

Ian went on to say that we had to get away from the farm for periods of at least one to two weeks. This time away allows you to rest, relax, and come back rejuvenated.

The very next day we combined all three herds into one large herd. We walked them down the gravel country road, stringing polywire to keep them off of people's property. (Bulls were removed to keep them from the heifers.)

The immediate work load was cut by two thirds!

Our rest periods increased two thirds!

The next evening Jan and I took a leisurely horseback ride around the farm instead of running from one farm to the next moving wires for three separate herds. I had moved the herd that morning so there was nothing to move that night.

From this day forward, all custom grazed cows will be added to our owned mob and grazed as one mob.

One of the comments Ian made was that humans tend to over complicate everything that we do. I can surely identify with that remark.

In the future we will grow more grass, better grass, have more animal impact, our soil microbes will explode, our ground litter accumulation will benefit, our water catchments will increase, and our labor has been slashed dramatically.

We now have a much better quality of life.

We have time to think, monitor results, and for leisure.

We no longer have the issue of getting burned out with something that we both have a tremendous amount of passion for and dearly love.

That one single decision to combine our herds has already paid huge dividends. Jan and I now have time for each other, wow what a wonderful life!

Thanks, Ian, for dropping the brick on our head!

Chapter 33
Let the Cows Do the Walking

It's amazing how we get settled into certain ways of thinking and develop a mindset that this is the way it has to be done.

I had a major paradigm broken when South African rancher, Ian Mitchel-Innes, suggested that we combine all three of our herds into one large mob.

My biggest hang-up about his idea was the hauling of cattle from one farm to the next when they ran out of grass.

Previously with three smaller separate herds we had been able to keep each herd on each grazing system without running out of grass. We were stocked light enough that we rarely ran out of grass.

I literally had nightmares thinking about the cost of hauling 250-500 head of cattle from one farm to the next depending on the growing season.

In the summer we are stocked at 500 head, but the winter stocking rate is only 250 head. The trucking bill was going to take a serious bite out of our pockets, not to mention the stress on us and the animals from hauling them.

Ian being the gracious guest that he was, came straight to the point.

"Cows have four legs and are born to walk.," he said. "Why do you want to go through the expense of hauling them when you could let them walk?"

Ian went on to say that we ought to charge people to participate in our cattle drive and make it a festive activity. He said to put on a weekend campout, bonfire, chuck wagon feed and charge people to bring their horses to participate in our cattle drive.

We got paid to move the cattle instead of paying someone to haul them. How can you do any better than that?

We started planning out our first cattle drive, although we did not do it exactly like Ian suggested. It is our future goal to put together a cattle drive like Ian suggested, but this will take some additional planning.

As the time neared to move our 250 head of cattle to our next grazing system, we started by contacting our neighbor who borders our biggest leased farm.

The back of our leased farm runs two miles north from the blacktop frontage road and is isolated at the very back property line.

This is where I started eyeballing our neighbor's farm that bordered the leased farm for a possible route out to the gravel road that takes us to the Judy Farm.

My neighbor had five cows and one bull on his farm for the winter. We called him up and asked him if we could build a temporary polywire lane across the end of his farm that would allow us to move our cattle herd out to the gravel road.

He had several concerns with us doing that. One concern was getting his cattle mixed up with ours, which he solved by suggesting that we could pen his cattle in a lot for 30 minutes while we brought the mob through his farm.

His other major concern was with our large concentrated herd that we would destroy his pastures and leave a mud lane on his farm.

We had recently gotten three inches of ice followed by two inches of rain, which made the pastures pretty soft. I promised him that we would not move the cattle across his farm unless the ground was frozen hard.

Then I prayed for a good cold snap to harden up the ground.

Our targeted day of the upcoming cattle drive was a

Sunday morning. This is when most people sleep in and the roads have minimal vehicle traffic on them.

By using our grazing chart, we knew exactly how much forage and time we had left on the large farm before we ran out.

A concern I had was that the ground would not be frozen hard enough to move the cattle across our neighbor's land when we needed to move them. Once we arrived at that date, what were we going to do with the cattle that were out of grass? For an insurance policy, we hauled in enough hay and dropped it off in some paddocks behind the cattle to help hold them over until the ground froze.

The Saturday before the scheduled cattle drive, we built lanes across my neighbor's farm. We moved the cattle into the last strip of stockpiled grass that remained on our leased farm and held our breath for a cold nighttime temperature.

Sunday morning we got up before daylight and walked out in our yard. The ground had frozen hard during the night.

We jumped on the ATV with enough polywire reels and temporary posts to fence off the unfenced road areas leading to our farm.

As we drove down our driveway we opened the gate and tied a white ribbon across our driveway, which would divert the mob into the new stockpiled pasture. Once we got out to the gravel road, we tied a white ribbon onto our mailbox and just laid the ribbon on the road leading over to our drive-way. I did not want to tie the ribbon tight across the public road in case a car came by, but I wanted it ready so it could be secured very quickly.

As we worked our way down the gravel road, we ran one strand of polywire and secured it to brush and temporary posts where the wire needed support. We had three different unfenced yards that we had to pass and several brushy areas that had no fence.

Next, we went up the temporary lane we had built the previous evening and moved our neighbor's cattle into a holding pen.

None of our polywire fence or lane was energized with electricity. Our cattle are hot-wire broken very well, and basically we cannot push a cow through one of our wires.

As luck would have it that morning, the cows were not ready to move. We had given them a much larger strip of grass the day before because we knew that we were not going to be back on this farm until May. The cattle were all filled up and lying around with smiles on their faces.

Jan and I got out our trusty white ribbon and got behind them. With the white poly tape reel we moved the herd up to the opened gate and off they went like a bunch of school kids.

I got in front of the herd with the ATV and Jan was on foot bringing up the rear, keeping the stragglers caught up.

We shot through the neighbor's farm in about 15 minutes and hit Devils Washboard Road. That is the name of the road we live on, and it is appropriately named.

The whole length of the road is one large hill followed by another larger hill. Once I came to Devils Washboard, I shut off my ATV and listened for traffic. There were no vehicles coming from either direction. This was the window of opportunity that we were looking for.

I tied a white ribbon across the road by the gate to keep the animals from going the opposite direction that we wanted them to go. The cattle were right on my heels when I took off down Devils Washboard on the ATV.

The herd spread out in a long line. They were shooting down one hill and charging up the next hill.

Life was good, the cattle were happy to be going somewhere, everything was going great! We were near the Judy Farm entrance when we encountered an unforeseen obstacle.

Right on top of the last big hill overlooking our driveway were four big Alaskan Huskies. They saw the cows coming and came charging out of our neighbor's yard.

Thankfully, the yard was fenced all the way to the edge of the road, which physically kept the dogs out of the cattle, but the cattle would not move one inch further down the road due to the barking, slobbering, pack of dogs.

The long strung out row of cattle were now one huge bunch packed on top of the hill. The mob on the hill looked like a balloon filling up with water that was going to bust at the seams at any minute!

The whole mob turned and took off running back toward Jan.

Jan had not made it out to Devils Washboard road yet. She was just bringing the last stragglers out to Devils Washboard. She said it sounded like a stampede of buffalo coming at her down the temporary lane.

Jan took off running backward and shut the gate behind her before the mob reached her, and bravely held her ground. I came back to the rear of the herd on the ATV and helped her move them back up to Devils Washboard.

With Jan and I both at the rear of the herd we got them past the barking dogs and within five minutes they were on the Judy Farm contentedly eating stockpiled grass.

Hindsight is always 20/20 and we should have first knocked on the neighbors' door and asked them to pen their dogs for five minutes. That is definitely on our "To Do List" next time we move cattle by this place.

We took the herd up the temporary lanes, released our neighbor's cows from the holding pen and were done. I took Jan out for breakfast that morning at the local restaurant. We were sipping coffee and eating flapjacks and sausage by 9 a.m.

We had moved 250 head of cattle three miles and it had not cost us anything because we had let them walk. The headache of sorting and loading was eliminated as well. Our cattle were not stressed, and we had not been stressed except for the barking dogs incident.

We saved over $1000 in trucking that morning, a pretty good return for an hour's worth of cattle walking.

As we returned from eating breakfast, it was a wonderful sight to come down our driveway and be able to see the herd chomping down on fresh stockpiled forage.

We had six weeks of good stockpiled forage on this farm. Once this farm was grazed off, we planned to move on to our last stockpiled farm, which would last the herd to April 10th.

Here are a couple of things that we will do different next time.

First, we will visit all neighbors with dogs and kindly ask them to pen their dogs for 30 minutes or until our herd has walked by.

We will not take the herd a long distance on roads without at least two blockers in the rear of the herd.

The blockers will have a 60-foot piece of white ribbon to hold between them in case the whole herd turns back on them.

Also, it would not hurt to have a person at each end of the public road blocking vehicles from entering the herd while they are on the road.

At the current cost of diesel fuel and hauling charges, we are having fun letting our cattle walk. It is good for the cattle and good for our pocketbook.

As we do this mob moving and grow in confidence, it will become great FUN! When this happens, Ian seriously urged us to either charge people to help us or use this occasion to get "social and emotional credits."

Quoting Ian, "What a way to repay favors and get people involved and interested in what you are doing!"

Heck, I can see in the future that we may invite all of the neighbors along our Devils Washboard road to take part in our cattle drive. They could do several valuable operations for us.

One, would be to act as road patrols at each end of the road while we move the mob down the road. We could have a person opening and closing gates as we move the mob toward their destination.

As a reward for participating in our cattle drive they would be invited to a grassfed beef cookout that evening. I can't think of a better way to get your neighbors educated on what you are doing than by having them be a part of the process.

You can explain the importance of having one large mob of cattle impacting an area, then letting it rest and re-grow before re-grazing it. I'm getting excited. Let's have a neighborhood cattle drive and reward them with a nice beef cookout. Talk about building a strong connected community, this may be it!

The neat thing about this whole concept is that we have taken a liability (hired hauling) and turned it into an asset in the form of a potentially strong community building tool, which is sustainable.

Chapter 34
Grazing Tips from Ian Mitchell-Innes

When Ian Mitchell-Innes spent several days with us in the fall of 2007, I was like a little kid in the candy store. Ian never left me feeling like I asked a stupid question. He answered every question with extreme detail and shared a treasure chest of information learned from his years of Holistic Planned Grazing while ranching in South Africa.

I heard Allan Savory (founder of Holistic Management) say that Ian was the first rancher in the world to get all the pieces of Holistic Planned Grazing in the correct order. Many others before him had failed to recognize all the important building blocks of a successful ranching operation.

While walking around our farms, one thing I immediately recognized was how aware Ian was of his surroundings. He would comment about something in the pasture that was very obvious to him, while I was oblivious to it. I wish I could have spent several weeks quizzing him about things, but am very thankful for the time we spent together.

Now, I am going to share my notes from Ian with you.

The Importance of Carbon:
Every unit of carbon holds 12 units of water. You build carbon by concentrating on adding all the organic matter you can on the soil surface and trampling it in the ground with livestock. This is where High Density Grazing really shines. The high animal density gets the live material laid flat on the ground as litter.

By adding a higher volume carbon load you feed your soil's microbes and earthworms.

The more carbon you have in your soils the longer the quality stays in your grasses. Plants that are healthy are not fighting to survive. This reduces the need for the plant to immediately send up a seed head, and gives you more green growing leaves. The sole purpose of a plant is to survive and produce seed to replace itself for the next growing season.

Base your moves on giving your animals the best selection they can easily eat and letting what they don't eat get trampled on the ground as litter. The improvement of your soils is related to the amount of litter (carbon) you can get on the soil surface. Carbon on the soil surface serves as the skin of the soil. You must be able to get solar energy through the soil surface.

When an animal grazes a plant, carbohydrates (sugar or energy) shift down into the plant roots. This is the only way that energy from the sun can penetrate the soil surface and feed the microbes and other living organisms in the soil.

Microbes:

If takes five days for microbes to change from one generation to the next. If animals are grazing good pastures and are moved regularly, the microbes needed to digest that quality of pasture will develop in the cow's rumen. Then, if you pressure the animals and leave them in a paddock for a longer period (maybe to create landscape) after five days the rumen microbes start to change to those microbes that can cope with the pasture that is lower in nutrition.

Say you leave them for another five days so that all the rumen microbes that deal with the good pasture have died off, and you then move them to another good pasture. The animals will not get the benefit of the good pasture for five days, because the rumen microbes have to change back to **good** pasture microbes.

We are not working with livestock or grass, but microbes in the stomachs of animals and in the soil. Optimize both and you will be working with everything that is God given and spending nothing except management time. I now concentrate on being a microbe farmer instead of a grass farmer.

Forage Topics:

Properly grazed pastures always grow back faster than mechanically mowed pastures. It is the pull, action whip of the cow grazing grass that triggers a microbial explosion under the ground. I had never heard this before, but I have noticed how slowly hayfields grew back compared to pastures that were grazed off.

The top 1/3 of the plant is where the energy is stored. The middle 1/3 of the plant is where the fiber is stored. The bottom 1/3 of the plant is where the protein is stored.

Oxidizing grass (grass left standing upright by low stocking density) in your grazed paddocks can be addressed with High Density Grazing. Once you increase your stocking density, this grass gets trampled or eaten. I had this problem in many areas of our pastures with our old grazing system. Since switching to High Density Grazing we no longer have areas that are avoided by animals. The utilization rate of your pastures skyrockets when all the grass is either eaten or trampled for microbe food with each grazing pass.

Burning reduces production, and kills plant roots three feet down. I have already told you how I feel about burning grasses. If you want to add carbon to your soils, burning will never equal animals treading the dead matter onto the soil surface.

If you have a poor field and a fertile field next to each other, graze the fertile field first then rotate to the poor field. What Ian is saying is that by grazing in this order the livestock are transferring nutrients and seeds from the fertile field and helping build up the poor field for free.

Take the soil sample through the center of the grass plant's tuft. This is the interface between the rhizomes and microbes.

With High Density Grazing nothing is ever wasted. If you only graze 20% and get the other 80% effectively on the ground, you are just banking your money. It's like buying a 50% share in a fertilizer factory, only better because it is sustainable. Ranching becomes highly profitable.

If you do not allot enough forage, some cattle just stop eating after being bullied enough by other dominant cows, and animal performance plummets.

Building Your Herd:

Buy two young heifers for the price of one bred cow. Graze these heifers through your management system and keep the ones that perform well, sell the ones that don't.

The bull that works at 15 months is a desirable trait. How do we know which bull is the best? Let nature sort out the best bull. Ian only uses a bull once. They do not castrate in Africa.

22 days is the ultimate target for a bull to be in the herd. Any cow that gets bred outside this window will eventually end up falling out of the herd (not rebreed, etc).

Take your time getting to the goal of the 22 day bull exposure process. Each successive year, shorten up the exposure time you give the bull to the herd.

Big heifers are freaks of nature. Do not use them as breeding stock in your herd.

If the ph of a cow is below 7, she will not conceive. Ian immerses a strip of ph paper in the urine puddle of a cow to check her ph level.

Anything that is out of sync with nature gets culled. Big fat animals and skinny ones get caught and eaten by predators. Nothing out of sync with nature survives. In modern animal agriculture we continue to propagate these animals and their offspring with terrible consequences. Animals that would have perished in nature get to survive and pass on their inferior genetics to the detriment of your herd. Ian's present average cow weight is 1144 lbs, his target is 880 lbs.

When you walk into a herd of animals to move them, target animals that look stressed. Observe from the left side standing behind them. If you see a sunken triangle between the hipbone, rib and backbone, it's a sign of a hungry animal.

In the autumn take off the bottom 10% of your cows that do not fit your management system. Develop a hamburger market for these animals.

Stocking Rate:

Stocking rate determines selection, which determines performance (money). The trick is to increase stocking density first and keep the stocking rate the same until you see how much more grass you will grow. Many people increase their stocking rate and stocking density at the same time, which ends in a huge failure.

Diseases only come into play with high densities if animals are left in an area too long and become stressed. Under stress and bad nutrition the ph of the rumen will change and the immune system of the animal will fail.

Diseases disappear with Holistic Planned Grazing because animals are moved onto fresh clean pasture and do not come back until the area has been naturally sterilized by the sun. Most diseases we have today are a result of management.

You can not get fully stocked if you ignore animal performance (stocking rate).

When paddocks get away from you, do not mow them. Keep them as a fodder bank, which can be used in dry periods or during the winter. This is much cheaper than mowing or baling.

Animal Performance:

Two months prior to calving, concentrate solely on animal performance at all costs, up until two weeks before removing the bull from the cowherd.

In other words, make sure your daily grazing allotments give the livestock the best they can eat. 80% of the calf's growth takes place in the last two months of pregnancy.

Important Note: Do not be in the landscaping mode during this time. When you are in the landscaping mode, you are focused on building soils not so much animal performance. Landscaping mode is usually done with dry bred cows that have low nutritional requirements. Animal performance grazing is focused solely on getting the very best forage in them and all they can eat.

The Importance of Thinking Time:

Whenever making difficult decisions, put them off for a while and think about them under the shade of a tree. You may find out that you did not need to do them after all.

When facing difficult situations, back off a bit and think what would happen before man arrived with the firearm.

Attitude leads to gratitude. Be the person who you would like to meet that day. Ian keeps a stone in his pocket to remind himself of this daily.

Have the right attitude. What do you want to achieve? Why? Why? Why?

Quantify the asset that we are working with. How are you going to position yourself for this opportunity?

Go through the Holistic Goal Setting and Decision Making process.

Do a marginal reaction, (time and money spent). When comparing two enterprises, you need to look at the limiting factors of each.

Synergy, trust, companionship, help, - we all desperately need all of these to flourish.

When are you going to do it? Where are you going to do it? How are you going to do it?

Plan. Plan. Plan. It is about understanding what you want to do.

And Finally:

Use a sports whistle to move your cattle. This allows

anybody to move your cattle (you can take a vacation). It also gets more excitement in your herd, which aids in animal impact on your pastures.

Animals were made to walk to water, don't spend tons of money on expensive water points.

Remain flexible, monitor animal performance, monitor animal behavior, monitor animal fill, remain flexible, ENJOY YOURSELF!

Chapter 35
Landscaping versus Animal Performance

Using High Density Grazing as a tool, the term "Landscaping" means we are concentrating on building the soils by trampling forage, building litter, removing rank forage, eating brush, using animal impact, etc.

"Animal performance" is the method used to get the most highest quality nutritional forage through your animal when it needs it the most.

It is extremely important to know the difference between these two practices and when to apply them when you are using Holistic Planned Grazing.

When we started using High Density Grazing the results from increasing our animal density were absolutely amazing. I was so focused on landscaping because of our immediate grass improvement that animal performance was neglected at a key time. This time period was two months before the cows started calving.

I was moving them every day, but was not giving them a large enough selection of forage. Eighty percent of the calf's growth comes in the last two months of the cow's pregnancy. If the cow does not get everything she needs in this time period, her calf will suffer.

We had some pinkeye in our new calf crop that probably could have been prevented if we had been more focused on animal performance instead of landscaping. I believe the cows were not able to pass all the good antibodies along to their calves due to the lack of forage selection.

It seems that once a calf has a health issue, we immediately start throwing medicine at it without asking what caused the calf to get sick in the first place. A calf usually gets sick because it is not getting the proper antibodies from its mother's milk. Why does the cow lack the proper antibodies in her milk? Usually because our management practices caused her to limit her intake of quality forage.

She needs to be gaining weight every day, increasing her body condition score. Cows usually lose some weight over the winter by taking their calf through the winter on the fat on their backs. Ideally her body condition score should be 6.5 at time of calving, which will allow her to put everything into her new unborn calf that it needs to be healthy when it is born.

A 6.5 body condition score at calving time will also ensure that the cow will breed back on schedule for next year's calf crop. If your cows are thin at calving time, they will be late breeding back, and some will not breed back at all. If they do not breed back, you have a huge loss to deal with by getting rid of the open cow. Open cows can give you some sleepless nights unless you have a good hamburger clientele to market them through.

We calve in June and July, so that means we need to really focus on animal performance during the two-month period before they calve. This means we will give our mob a larger area with each paddock move to give them a higher selection.

When we combined yearlings, cow/calf pairs and bred heifers into one large mob in the fall of 2007 it was like we got our life back. Moving one herd a day instead of three herds was like I was on vacation. But, it also made us focus more on animal performance because we had the coming two-year-old grassfed beeves mixed in with the mob, and they required higher quality forage. We have been more focused on giving the whole mob a higher selectivity with each paddock area that is erected.

The spring of 2008 was a challenge to manage for

animal performance. We had coming two-year-old grassfed beeves to be finished on grass, weaned yearlings, June-calving bred heifers, older June-calving cows, and custom grazed dry bred cows all in one large mob. This required us to be extremely focused on animal performance during that period. The dry bred cows probably thought they were on vacation with the animal performance treatment they received that spring. The previous two years they had seen nothing but landscaping mode grazing due to their low nutritional requirements.

Ian has a great animal performance chart that shows the important dates for animal performance and landscaping mode.

The starting date for animal performance is two months before the cows calve and ends two weeks after you take the bulls from the herd at the end of breeding season.

After this period, the cow is bred, and the calf has some age on it and is eating grass. As soon as the calf starts eating grass, this removes some pressure from the cow and lowers her nutritional requirements. This is the time period for landscaping, (building soil) using your cowherd. The bred cow's nutritional requirements will be lower, which allows you to push her a little harder with landscaping chores - busting up rank forage, cleaning up weedy areas, removing brush - are some of the things that come to mind.

To summarize animal performance and landscaping mode keep these points in mind:

1. Never use landscaping mode the last two months of the cow's pregnancy.
2. Two weeks after the bulls are removed, landscaping mode can be used.
3. Animal performance is extremely important with finishing animals.
4. Focus on achieving a cow body condition score of 6.5 at calving date.

Genetics and Grass-Finished Beef

Chapter 36
Grain versus Grass in Your Cow Herd

I never realized how much difference there is in groups of cattle until we started custom grazing. Each group performs differently.

A large majority of our cattle today have been developed by concentrating on how the cattle perform on grain in our nation's feedlots. This has been very detrimental to our nation's cow herd. Around 1950, our country developed synthetic fertilizer and found out they could grow an abundance of corn by applying purchased fertilizer to their fields. This set in motion a gluttony of available corn.

Today 70-80% of our nation's corn is fed through a feedlot animal. This has been a disaster for the cow/calf producer trying to make a living out on the farm with a group of beef animals.

The cows have been bred to such a large frame, which is what the feedlots wanted, that they have trouble producing a calf each year on grass. The larger the frame, the more feed it takes to maintain her body condition and breed back each year. A cow that does not breed back is a huge economical loss. In Missouri, they are usually put into a fall herd simply because they did not get bred for the spring calf.

You lose half a year of production from that cow every time you move a cow back to the next calving period. If we could raise a cow the size of an elephant, the feedlots would love it, because they could feed it more corn. They are in business to sell corn, and they get paid every day they can feed corn. So if they get a particular bunch of cattle straightened out, they can feed corn for a long time to large framed cattle simply because they have more area to put the weight. They do not want an animal that fattens quickly. This shortens the time that they can feed corn to them.

This all may change now that the ethanol boom has started. Ethanol plant demand will suck up a bunch of the United States' corn. This may be a good thing for us grass farmers.

As corn prices go up, our grass gain will be worth more simply because it is cheaper to put gain on with grass than it is with five-dollar-a-bushel corn. Calf prices may go down when corn prices rise, but we can still make a nice profit if we concentrate on our grass.

A smaller cow, let's say 1000 to 1200 lbs, will always wean a higher percentage of her body weight than a 1400 to 1600 lb cow every time. Do not get caught up in the bigger is better mind set, or your grazing operation will go down the drain. If we calculate 2.5% of a cow's body weight for daily consumption and compare a 1000 lb cow with a 1500 lb cow the numbers are staggering.

You can graze 100 one-thousand-pound cows on the same amount of grass that it takes for 67 fifteen-hundred-pound cows. Now I'm not a rocket scientist, but that is 33 more calves to sell at weaning time from the 1000 lb cows. Those extra 33 calves is where you walk to the bank smiling.

Today there are very few cattle out there that can support themselves solely on grass, because this ability has been bred out of them for the last 50 years. Cows grew for centuries on solely on grass and they did fine. Cows are herbivores, not grain-a-vores. The grain actually upsets their rumen, which is designed for grass. If you want to make money in the cattle business in good and bad times, the only sustainable operation is a grass based operation.

No matter how cheap grain is, it can never compete with grass. You do all the work of raising and caring for that calf. Why not put an extra 400 lbs of grass gain on them and capture that profit for your own operation? We focus on getting as much grass gain as we can on them from free sunlight.

For the beginner just getting started in grazing, developing your own grass genetic herd should be a goal that you can work toward. Develop your grazing skills by custom grazing other people's cattle. Save your money and develop your own grass genetic herd as you gain confidence and knowledge.

Chapter 37
Developing or Finding Grass-Genetic Cattle

So where does a person find some grass-genetic cattle?

The most economical method would be to attend local cow sales and look for small frame cows. These smaller frame cows will always be cheaper to buy than the monster cows that all the mainstream cow producers want.

These smaller cows will make some good candidates for starting an efficient easy keeping grass-genetic herd. When you're looking for these smaller frame cows, watch for cows that have a big gut for lots of grass capacity. Most cows today have had all the gut capacity bred out of them. We call them pencil gut cows because they will suck your pocketbook dry. A pencil gut cow has no grass storage capacity, which results in cows that eat a lot but still remain thin.

Breeds of cattle that we have commonly found to make good grass-genetic candidates are Red Devon, Southpoll, Murray Grey and Angus. There are many other breeds that may work just as well. I only listed the ones that we have used successfully.

If you buy some grass-genetic candidate cows that have been fed grain in their past history, just bring them home, give them solely grass and watch them. If they start losing weight by solely grazing grass after several months, sell them. It will take a month to get their rumen adjusted to a grass diet after consuming grain.

Cattle producers have been misled by input salesmen for so long about what cattle need and don't need.

All this does is take money out of your pocket and put it into theirs.

Most people still worm their mature cows every year. What a waste of money. I've never wormed our Southpoll herd and they perform fine. If a cow gets wormy, sell her. Don't let her propagate her poor genetics into your cow herd.

If a cow misses calving, cull her. There are no acceptable excuses for keeping her.

If a cow gives you a runt calf, cull her.

Bad attitude or skittish cows should not be tolerated. They will pass this same trait on to their offspring.

Cows that develop foot problems should be culled from the herd.

Cows that come through the winter with low body condition scores should be culled.

Cows that do not shed off early in the spring should be targeted for culling.

If a cow needs assistance calving, cull her from your herd.

You're probably thinking by now, if I cull for all those reasons, I won't have any cows left in my herd.

One option is to divide your herd into two groups. The elite group will make up your future herd. Keep the best offspring as replacements. Label the other group as calf makers and sell all their offspring.

Culling cows can be expensive, but you will never have an elite grass genetic cow herd unless you cull ruthlessly. Make that cow work for you, not you work for it.

If you put money in a savings account in the bank, you expect to draw interest on it every month of the year that it is in there. Look at a cow the same way. She has got to be able to suckle a calf 10 months, be growing one inside her, and breed back on schedule every year while maintaining a good body condition. These are the kinds of cows that you can build a profitable cattle business around.

The best time to pick the elite cows in your herd is late

winter, right before the grass starts growing. A really good grass genetic cow will get overly fat if you wean her calf at seven months strictly because she has no pressure being put on her from the calf. This same cow may have trouble calving the next spring because she is too fat from not having to work.

This late winter, early spring period is what separates the good cows from the mediocre cows. The good cows will winter well and not miss a beat. They will shed off early and pour weight on with the rush of spring grass before they calve.

I had the good fortune to learn the importance of grass genetics from Gearld Fry. He is the most knowledgeable grass-genetic cow fellow I know. He is committed to turning our nation's cow herd around so that folks back on the farm can make a decent living from their cows again. He has a lot of people who disagree with him. I'm not one of them.

Gearld studied Jans Bonsma, a cattle geneticist from Africa who developed the linear measuring method. He is passing this wealth of information on to anyone who wants to learn it. If you really want to learn the nuts and bolts of developing a good grass-genetic herd, you need to get in touch with Gearld. He has a web site called Bovine Engineering, and is part owner of a company called the Bakewell Reproduction Center.

Following convention, I was failing and knew there had to be a better way to develop a profitable cattle herd. My cows must take care of me or they don't have a place on my farm.

There is a great system for actually measuring your cattle that will tell you how good your grass cattle are. It is called linear measuring. It consists of a whole list of different measurements that you physically collect by using a tape measure. You need a head catch to hold the cow or bull in order to collect the data. You can not eyeball a cow and get these measurements. The tape measure does not lie.

One of the measurements consists of heart girth, which measures the circumference around the cow right behind the front legs. This measurement is critical. If your animal has a pinched heart girth she will be a high maintenance animal. Her internal organs will not have adequate capacity to perform at full performance. Most of our nation's cattle today are lacking adequate heart girth. The top line measurement is taken from the poll (top of head) to the rear end.

Ideally, the heart girth should be equal to the top line of the animal. If the heart girth is more, that is excellent. Our cows should look like boxes of meat supported by four fine-boned pegs. We have too much leg in our cattle today. The more open space under the cow's belly equals a higher maintenance cow.

For cattle to perform on grass, they need a lot of gut capacity to put the grass. The cows should have a big old butt on them. This is a feminine trait. It gives them more room to calve. A cow needs a big butt. Her udder should be flat, in line with the bottom of her belly, not hanging down, or worse, tilted back. Her hip bones should not be sticking up out of her meat. They should be smooth. The animal should have slick oily spots on each side of her lower neck. This shows a well operating thymus gland. You can see this by looking at them from each side. If you stand directly behind an animal, they should have a dark oily streak running right down the center of their backbone. This kind of animal is performing well in its environment.

When picking replacement heifers from your herd, you should always linear measure them first. Be careful not to pick the largest heifers. These are freaks, and they will increase the frame size on your cow herd. Your goal should be to have mature cows that average 1000 to 1200 lbs. These are the most efficient cows sizes you can have.

It truly is a wonderful feeling to go through an entire winter solely on stockpiled grass. The cows don't mind grazing in the winter, in fact, they prefer it. They are a lot healthier grazing than standing around a bale ring in the mud.

If you have to buy some grass-genetic cows, find a good source, bite the bullet, and buy several good cows. Start saving the best heifers and grow from there. We started with 22 pairs in 2003 and by the spring of 2008 we had over 150 head of breeding stock.

Good grass-genetic cows are as scarce as hen's teeth, but there are some out there. We may be selling some of our grass-genetic heifers in the future once we get our numbers up.

Chapter 38
Grass-Genetic Bulls

The bull is the foundation of your herd. A very good grass-genetic bull is very difficult to find. However, if you linebreed, you can eventually raise your own.

Linebreeding consists of breeding within the herd. By breeding within the herd you are building your gene pool. As long as you cull ruthlessly, your herd will get better with each generation. You should linear measure your bull candidates to see if they are suitable for being a herd sire. A lot of bulls are shaped more like cows than bulls.

The one measurement you will have a problem obtaining is the testicle circumference. Ideally, according to Gearld Fry, it needs to be 36 cm at two years of age.

Be careful with this measurement when looking at bulls. Some grainfed bulls will meet this measurement simply because of the fat that is built up in the testicles. I have watched several bull sales, and it seems the biggest fattest bulls bring the most money. This makes no sense to me. All you're doing is buying a pampered bull that will leave some of your cows open and end up looking like he is standing in a soup line. There are cheaper ways to get hamburger than grinding up sterile herd bulls.

To measure the testicles yourself, put your bull in a head catch and encircle the testicles with a string. Mark the point on the string that gives you the testicle diameter, then measure the string with a tape measure. Convert inches to centimeters and you have the measurement.

Never feed your bulls grain. It kills their live sperm count. By feeding grain you're building up fat in his testicles that will severely affect his breeding ability.

The testicles should have a shape like a football, not long and pointed. The problem with most bulls you buy is that they are grain fed from the time they are weaned and probably were creep fed before that. These bulls will fall to pieces when you place them out with a group of cows to breed. They simply can not maintain their body condition and breed cows.

A common practice is to place bulls in a pen and feed them grain and measure their daily weight gain on feed. The higher their daily gain, the more money they supposedly bring. This has absolutely nothing to do with getting your cows pregnant.

Your bull should have nothing but forage to grow on. When you turn him out with the cows he will be right at home with his preferred and adapted food - grass, and will not be walking around the pasture looking for the corn feeder.

These forage grown bulls have bred as many as 75-100 cows by themselves. The common number of cows bred with grain fed bulls is around 20-30 cows. This means you may need two to three times as many bulls to get the same number of cows bred as you would with one good forage grown bull.

In 2007 we purchased a two-year-old, two-frame, registered Red Angus bull from Pharo Cattle Company of Burlington, Colorado. We wanted to add more volume and lower the frame size of our cows. We hope this bull will give us heifers that fit this model and can be added to our cow herd.

This young bull is built like a tank. He looks like a box of meat on four short posts. He is extremely thick, has a big gut, short legs, slick hair, short thick neck, wide chest, and is 49" tall. He has never had a bite of grain, and always stays fat on grass while breeding cows.

Pharo Cattle Company is one of the few seedstock producers that develops their young bulls on forage alone. This is the only kind of bull you should ever buy. Kit Pharo started the company and now has 15-20 producers raising young bulls under his watchful eye.

We actually had a grain-fed bull go sterile one month after turning him out with the cows. He was pig fat from eating corn when we turned him out with the cows. He left the cows and stood in the shade of the trees all day. During this period there were cows riding each other, but we saw no sign of interest from the bull.

The bull turned meaner than a snake. If you got close to him, he would flat come after you in a hurry. He had me isolated one day with only a tree between us. When he charged me, I would circle to the opposite side of the tree. This cat and mouse around the tree went on for about 30 minutes, before the bull got tired.

I called the cattle owner and told him the bull was so ornery that we could not even get around him. He wanted to come out and see the bull for himself. He thought surely the bull wasn't as bad as that. We drove in the gate to the pasture and as luck would have it, there the bull was standing under a big cedar tree about 100 yards off.

The owner got out and stood by the truck looking at the bull. The bull never moved one inch, just stared up at us. Then the owner started walking toward him at an angle and stopped about 100 feet from him. The bull never moved.

About this time the owner got a call from nature and unzipped his pants to relieve himself. Well, right in the middle of relieving himself the bull immediately charged. The owner never had time to zip up, only to take off running down through the brush. I believe he screamed first. The last I saw of him for awhile was him running for his life down through the cedar grove with the mad bull in hot pursuit.

About 10 minutes later the owner came sneaking like a snake back up out of the brush, and the front of his pants were just covered with something other than water! Still breathless from running for his life, the owner got in the truck and said, "By God, that crazy bull will kill you. Call the butcher and shoot that darn thing."

Sometimes a person just has to see for himself to believe what you tell them.

Of course I dried up my tears from laughing before he got back to the truck. Do you know how hard it is to outrun a mad bull with your zipper down while you are in the middle of relieving yourself? I still can not help but chuckle when I recall the incident.

See how much fun you can have in this business? I highly recommend it.

Chapter 39
What to Do with Open Cows

I have made the serious mistake of giving an open heifer another chance to get pregnant.

I had bought this group of heifers that were supposed to have been bred. They were preg tested before I bought them, and were all in their second period.

When calving season arrived, they all calved except for three of them. Of course, it was the best three heifers in the group that didn't calve.

I was completely shocked that these good looking heifers were barren. You couldn't draw a better picture of a heifer that you would want in your cowherd as far as their looks went. They were deep bodied, thick across the chest, had big old butts, nice round bellies for storing lots of grass, a slick hair coat and were gentle as dogs.

Well, I started second guessing myself. What if the heifers had never been bred and the vet just missed them somehow? Maybe, if I gave them another chance they would just do fine. After all, where would I find heifers to replace those beauties?

You name it, I came up with a good excuse for them being barren, thinking next time would be different. I can not believe I was so stupid, but I just couldn't sell those heifers.

Folks, there was a good reason those heifers shined more than the others. They were getting a free ride, and no pressure was being put on them at all. They were on welfare. They were getting paid (eating our grass) and not doing anything to earn their pay.

Well, as Paul Harvey says, let's hear "The rest of the story."

Later that summer when we turned the bull in with the herd, those three heifers were fat as pigs, just right for butchering. But, old Greg is still in the mindset, "Wow, those are pretty cows!"

Well guess what? They did not settle again. I allowed them to eat our grass for one year and never got a single penny from them.

Let's do the quick math on this mistake. If I had custom grazed three cows for 12 months at $21 per month, that is $756 of total income, or in my case lost income. Mistakes like that can put you out of the cattle business real quick.

Take your lumps early. Sell the open cows. There is a reason they are open and you will not fix it with time. They belong in hamburger, not in your herd. The longer a cow stays open the harder it is to get her bred. There are tons of research out there to back this up.

Mark the date you want to calve, then turn in your bull to accommodate this time period. Some people expose their cows to the bull for 30 days and take him out. If a cow settles in 30 days, she stays in the herd. The ones that don't are sold. This is a great system, if you can afford the huge hit on open cows.

We leave our bulls in for 90 days, which cuts down some on culling open cows. I'm not saying this is the best system, but it works okay for us because we can use calves spaced out in age a little more for selling as grass-finished beeves. It gives us a larger marketing window.

The downside of our system is that we are keeping heifers and cows that are not as fertile as they could be. Culling cows gets very expensive, but it is even more expensive to hold onto open cows hoping that they will breed the next time.

You must have them preg tested 60 days after removing the bull. Anything that is open, sell it. I don't care how pretty they are! You can not fall in love with your cows or you will go broke.

I know folks who let the old cows die on their place. Their reasoning is, "Well, old Bessie raised me 10 good calves and she deserves a good place to die." These are folks who have allowed their emotions to affect their business. Whatever the old cow brings will be more valuable than a pile of bones resting in a draw somewhere. Save that money to buy a replacement cow.

It may take three to four old cows to buy one good young cow. That is still far better than letting them die on your place and receiving nothing. Some folks will argue by saying, "Man, you're cold blooded, selling old Bessie after she raised you all those calves."

Well, okay, I'm guilty. But, I have a young debt-free cow out there that is going to give me a calf next year and possibly many more calves in the future.

Chapter 40
Fine Tuned Grass-Genetic Machines

The grass-genetic cattle I have been describing can winter solely on grass. Larger frame cattle will suffer because their body is not made to fully utilize the stockpiled grass. To expect a 1400 lb cow to go through the winter suckling an early summer calf and keep its body weight up and calve on schedule would be like expecting a Volkswagen to beat a Porsche in a race.

With 1100 lb grass-genetic-efficient cows, they meter out just enough milk through the winter to keep the calf gaining. They don't milk heavily as this would be detrimental to their body condition. If you have a cow that just gives a little milk all winter, you have a calf that comes through the winter in great shape without any added expensive feed. This is a huge cost saver.

I've wintered cows in the past that were fed breeder cubes every day, had protein lick tubs available at all times, had all the stockpile and hay they could eat, but still lost weight over the winter and did not breed back. The cows simply could not take in enough dry forage to sustain their huge frames over the cold winter months.

I have to tell you about a common occurrence that I see a lot today with our stocker cattle.

We were custom grazing 185 heifers and steers for a fellow several years ago. I got them in March and was starting my rotation through my paddocks. By the end of May only two head out of the whole 185 had shed their winter hair. A lot of them would stand under shade trees all day to escape the heat.

A neighbor who shows steers at all the big stock shows stopped by and we were looking at the stockers. I made the comment, "Why won't those calves shed off?"

He kind of looked sheepishly down at the ground and replied, "Well, I guess the show calf industry might have a little bit to do with it." He explained that they intentionally breed long haired cows to long haired bulls so that the calves have long hair. This gives them more hair to shear in order to shape them up pretty for the sales and shows.

They actually owned a refrigerator room that they could keep at low temperatures so that they could place the show calves in that room to make them grow hair in the summer.

I was horrified and exclaimed, "How would you like to be exposed to these 90 degree days with a big long hairy black coat on?"

He commented that he needed to get home so that he could shear his State Champion heifer because she was getting ready to calve. This is a prime example of how broken our cattle industry is. Long hair is a sign of a hard keeping animal, not one you want to propagate. So, whatever you do, don't go out and buy the local champion heifer and start your herd from that.

I know I've probably just made a lot people mad who show cattle, so be it. The hard truth is these cattle have no place on a profitable, sustainable beef operation. Well, I got that off my chest. Let's move on.

Red hided animals are superior in heat tolerance compared to black hided animals. They did a study of black hided animals versus red hided animals in Gillette, Wyoming. They measured the surface temperature of both. The temperature that day was 90 degrees F. The red hided animal was 92 degrees F and the black hided animal was 110 degrees F.

I know right now the trend in the commodity beef market is, "Black is beautiful." I look forward to the day this changes. The cattle will be a lot more comfortable and perform better when they are red.

A black hided beef animal will actually lose weight when subjected to that kind of temperature for any lengthy period. They will also have more trouble breeding back strictly because of the heat. On a 95 degree F day I've actually seen our Southpolls lying down right on top of a hill in the full sun. They were not suffering at all. There were shade trees within 50 feet of them that they never used. They are red hided and bred for heat tolerance.

Another advantage of the oily hide that I mentioned earlier is this really repels flies. If you look through your cow herd, you will notice more flies on some cows than on others. Ever wonder why this is? The animals that have less flies have more oily hair and are performing at their absolute peak. Flies want nothing to do with these kinds of hides.

The Southpoll I mentioned earlier was developed by Teddy Gentry from Fort Payne, Alabama. Teddy got tired of trying to find cattle that would perform solely on Alabama heat and fescue. So he developed his own breed by crossing four breeds - the Hereford, Red Angus, Barzona, and Senepol. What he came up with after years of work was a four-way cross, very efficient animal that performs well in heat and fescue.

Teddy culled ruthlessly to get the breed to the point it is today. I remember the first time I saw them on grass at his farm, I about passed out. Never had I seen such a herd of slick, oily hided red cows and calves in my life. I didn't see a cull in the whole bunch. I feel very fortunate that Gearld Fry introduced me to Teddy and his Southpolls.

We purchased 22 cow/calf pairs and a bull. This is the base of our herd today. Teddy saved us years of developing a grass efficient herd. From the very first moment I got the cows, they have performed better than any cattle that I had previously grazed.

I remember worrying about the cold winter months that lay ahead. How would those Alabama cows handle our bitter winters and snow? Our fears were put to rest that first winter.

We got our first 10" snow. I drove back to the stockpiled field and they met me with grass hanging out of their mouths. They burrowed right through the snow and harvested the stockpiled forage.

I also worried about wintering the calves on the cows. All my neighbors had weaned their calves months earlier and were feeding them grain daily. Again, the cows did not disappoint. They wintered the calves and kept good flesh through the cold months. I had never seen cows that could do this.

Teddy has his own grassfed meat business and concentrates on raising tender meat solely on grass. He had a goal, and stuck with it until he had a very good herd of cows. I've never met anybody who seeks out data more than him and knows what to do with it when he gets it. He sent cows to town that most of us would have gladly put in our herd.

Teddy doesn't have to do this to make a living either. He is one of the four members of the singing group, "Alabama." Here is a guy who has done very well in the music business, but his real passion is reviving rural America. You can't help but admire a fellow like this. Keep up the great work, Teddy.

Chapter 41
The Marbling Issue

A misconception being taught as fact today is that cattle finished on grass do not marble as well as cattle finished on grain.

I'm here to tell you the truth. Cattle will finish on grass, but they have to be the right kind of cattle and be grazed on the right kind of grass.

The huge frame cattle that America's pastures are full of will not finish on grass. They have such a huge frame that they would be have to be grown for three or four years before they could marble correctly.

Smaller framed cattle can finish in two growing seasons on grass. Some very early maturing cattle with the right genetics will finish sooner than this.

One of the reasons grassfed beef got the slogan attached to it, "Tough as a boot sole," was that the animals were not finished. People were harvesting them before they got properly marbled. This gave grassfed meat a bad reputation to the general public, and bad news travels twice as fast as good news.

Genetics has a lot to do with tenderness of the meat. With linebreeding you can build tenderness into your herd by breeding within the herd. If you keep bringing in outside genetics, you are introducing constant unknowns, and you will probably never have a consistently tender product.

When finishing calves on grass, don't make them take down a pasture. Let them just top graze the candy, the very best. They are gaining at a very high rate when they are allowed to selectively graze. The closer to harvest, the more important this becomes. If the field looks mowed when you move them, you left them too long and their weight gain suffered.

You must have 30-60% legumes in your pasture for optimum gains with calves. Winter and early spring annuals will help you through the less optimum growing season. Some people give them all the quality alfalfa hay they can eat through the winter. This keeps them on a steady rate of gain. You can not allow a calf to drift backwards in weight gain as this builds toughness in the meat.

Recently I visited Jon and Wendy Taggart of Burgandy Beef in Dallas, Texas. They are selling 100% grassfed heifer beef and doing a fine job of it. They built their own state of the art locker plant, and everything is first rate.

I was super impressed with their meat. I purchased some sirloin steaks for dinner that night. We were staying with one of our landowners who lives in Dallas and he had thawed out some buffalo steaks for supper that night. I told him that I wanted to grill these grassfed steaks and compare them to the grain-fed sirloin buffalo steaks. He was a little bit skeptical that they would measure up to the buffalo steaks.

We treated them both the same, side by side on the grill. His kids were home from college and were anxious to try them. At the end of the meal the grassfed steaks were licked to the bone. They were absolutely awesome. Juicy and tender every bite! You could cut them with a regular table knife. We had a whole plate of buffalo steaks left.

Way to go, Jon and Wendy. You have an awesome product.

I know it has taken them a lot of time and hard work to get to where they are, but from what I witnessed, it is sure paying off for them. If you can recreate what the Taggerts are doing you can build a fine customer base. The meat sells itself.

Chapter 42
The Judy Grass-finished Meat System

We started our own grass-finished meat business several years ago to add more value to our animals.

When you sell to the commodity market you receive whatever they offer you, and you have no control over the price. We went through all the work of raising and caring for our animals then just took whatever price was offered. We would complain about the price received. It always seemed low after they took out all the sales commission, vet, feed, hauling, and other charges.

Selling to the commodity market does not reward you for your extra good quality, safe, wholesome meat. Once the animals enter the sale ring, it is just an animal, regardless of how it was raised.

However, we still sell on the commodity market because our grassfed business is so new that we have not built up the volume of customers to direct market everything we raise. But after numerous direct marketing sales, we have learned the importance of owning your own customers. When you direct market your products, you are insulated from commodity market price swings. When you own the customers, you set the price, not someone else.

I will cover how we raise our grass-finished beef first.

All of our grass-finished beef comes from our own herd of Southpoll cattle. Most of our calves are born from May to July. No winter calving is done on our operation. Cows must be eating green grass before they calve. This gives the cow everything she needs to raise the calf and breed back without losing weight. There is no supplement ever fed to our cattle, only pasture. We let the cows winter the calf on the teat, which keeps the calf in good condition through the winter.

The cows that struggle to keep weight on with this management process are sold. The calves are weaned April first using across-the-fence weaning. Basically, with this method, the cows and calves can see and touch each other through the fence that separates them. This makes them feel like they are still with mom. By that late date the calves are pretty much weaned by their mothers. It is basically just an emotional bond you are breaking when they are separated. After three days, the weaned calves have forgotten about their mothers and are grazing.

It is important to have some nice grass pasture set aside for this three day period. The calves have been grazing all winter. So during this three day period if you give them some nice pasture strips to graze, they forget about mom. After the third day you will hear little, if any bawling. Some cows handle it worse than the calves. They will stand by the fence and bellow while the calves are contentedly grazing grass. You must plan where you are going to wean them and have pasture growth available for that time period.

We have a five acre weaning pen that has a gravity flow water tank on one end of the paddock and the handling and loading facilities on the other end. It works very nice to simply walk the calves up and load them. We have not doctored a calf during the weaning process since we started leaving the calves on the cows through the winter. There is no stress on the calves. They are nine to eleven months old. This removes all the stress of being ripped off the teat at six months of age. I would never go back to early weaning calves.

These yearlings are sorted through, and the best females are selected for breeding stock. The bull calves are clamped with a Callicrate bander, which is a wonderful tool to use. The testicles just fall off in three to four weeks, and the calves never miss a beat. There's no blood, no flies, no mess.

In 2007, we grazed the entire yearling herd as one mob in daily or twice daily moves on some of our best pastures. We used the dry cow mob that year to condition the pastures for the yearlings.

The dry cow mob eats everything, which results in very tender new forage re-growing. It's perfect pasture for yearlings to put on weight. With this system our yearlings gained two to three pounds per day from mid April to the end of June. When the July-August heat hit, their gains fell off to one to one-and-a-half pounds per day, which was still pretty good considering the elements they were exposed to.

There are several advantages to raising your own grass-finished yearlings.

First of all, you can control the genetics.

Second, the yearlings are schooled to your daily moves right from the start by being with their mothers for the last year.

Third, you can select for gentle stock, which makes handling them a joy. Gentle livestock gain better and produce a more tender meat product. If an animal is stressed all the time, convinced that someone is out to get him, the animal will not be something you want on your place. Sell it and its mother. It is a genetic trait that you can control by selection.

Fourth, having your own yearling herd allows you to pick out the best for your own meat marketing program.

These yearlings are wintered the second winter on stockpiled forage and strip grazed through the winter. When spring grass starts growing we move them daily to large pad-docks where they have high selectivity. These paddocks have 50 to 70% red clover and alfalfa in them. The grasses consist of bluegrass, orchardgrass, brome, timothy, fescue, Reeds canary grass, gamagrass, and redtop. The steers only graze the very tops of the plants.

We move them daily regardless of how much forage is left. They may look like they are full when you open the gate to move them, but they will waddle through and start eating the fresh exposed paddock. They must have the very best forage every day and all of it they want.

These calves have a harvest date of June-July. The last 90 days of their lives is the most important to having a premium grass-finished product. We weighed the steers right at the beginning of May and again at finish. They averaged 2.4 lbs per day for the whole 90 days. That's 216 lbs of weight put on with a premium grass diet. You tell me how fossil fuel, corn-finished animals can compete with that! With increasing fossil fuel prices, we will be in the driver's seat.

The animals should waddle when they walk away from you. Watch the tail head for fat and wrinkles to accumulate. This is a sign that they are getting close to finishing.

I will never forget our first trip to our locker plant to visit the butcher after he processed our first load of steers. We asked to go back to the cooler to look at the carcasses hanging. We wanted to inspect them for marbling, fat cover, etc. The butcher sent us back with one of his meat cutters to show us the carcasses. He asked what our name was and what we brought in.

I told him we had brought in some grass-finished steers. He said, "Those will be easy to find." He walked up and down the rows and finally came back to us with a puzzled look. He said there were no grass-finished steers in there.

I informed him that there definitely was, and he should look again.

He commented, "Let me show you what a grassfed carcass looks like." He walked us up to an elk that had been butchered the day before. "That is what a grassfed carcass looks like!"

Hanging right beside the elk was a whole row of beautiful carcasses with a nice half to three-quarter inch backfat cover, and they had our name attached to every one of them. He cocked his head and said, "Those animals were fed grain. I have never seen grass-finished beeves with that much fat cover and marbling."

I'm not sure if the guy was ever convinced that we were not pulling his chain. I tried to explain the difference between grassfed and grass-finished to no avail. He was absolutely convinced that we had poured the corn to them.

His last comment was, "You can not make them marble on grass."

Obviously he was mistaken, you can get marbling on grass. It has to be using the right kind of grass and management. The animals must be mature before the marbling is laid down in the meat. An immature animal does not have marbling. An animal that was restricted in its diet will not have adequate marbling either. An animal that does not have access to lush new vegetative legume leaves will have trouble marbling as well.

The size of most of the cattle in the United States today is not suited for grass-finishing. In Argentina they finish their steers at 800-900 lbs, which is an ideal carcass to finish on grass. Maybe some day we can have those size animals on our farms, but right now we have to search for what is the next best.

One disadvantage to getting your cattle too small is that the commodity market will discount you heavily for a smaller framed feeder when you sell them. So maybe, for now we should try and stay somewhere in between large and small, especially if you raise more than you can direct market. If you can direct market everything you raise, go for the smaller ones. They are much more efficient.

Most of our cows range from 1000 to 1300 pounds. I don't want to go under 1000 lbs because we are still selling a lot of our production to the commodity market. This is one of our weak links that we are working on. It would be great to direct market every beef that we raised and in time we will reach that goal.

Since combining our herds and using High Density Grazing, we are trying a different grazing strategy for grass finishing our two-year-old beeves and our 10-month-old weaned calves in 2008. The first thing is to put the weaned 10-month-old calves back into the large mob as soon as the cows calve. By letting the cows calve first, this allows two months for the weaned calves to forget about sucking their mother. The coming two-year-old finishing beeves will also be grazed with the large mob in the spring. We will have everything in one large mob instead of three herds, which would normally consist of the cow herd, weaned calf herd, finishing two-year old herd.

One large herd will give us much better animal impact, longer rest periods, lower labor requirements, and most importantly a higher quality of life.

This new grazing strategy is going to be a challenge to manage. We are really going to have to stay focused on animal performance. With the younger calves and grass-finished beeves running with the mature cowherd we will have to give them a higher selectivity to keep the calves gaining well.

Chapter 43
The Commodity Side of Raising Yearlings

All the yearlings that do not go into our grass-finished meat or seedstock program are kept and grazed for the commodity market.

We target 800 lbs for a commodity marketing weight on our steers, and 750 lbs for our heifers. We feel like we might as well be getting the profit from the extra weight that we can put on our yearlings. If we sell them as 400 lb weaned calves versus 800 lb yearlings, we have missed a golden opportunity of putting weight on a straightened out, healthy, good grass-genetic calf.

If we can average 1.2 lbs a day gain on our commodity home-raised yearlings, that is 438 lbs for the year. The present pricing trend for heavy feeders in the commodity market is very good due to high corn prices. Heavier feeders require less feed to get them finished in the feedlot. Using 2008 feeder calf prices, a 400 lb steer is bringing $1.25 a pound, which equals $500. An 800 lb steer today is selling for $1.10 a pound, which equals $880. That is an extra $380 you can gross from each yearling by keeping it and grazing it on cheap grass.

Granted, you will have more time and expenses invested by grazing them longer, but our leased grass farms are an excellent tool to capture this extra $380 very economically.

Another advantage to having your own raised yearlings is that you are rewarded 100% by how well they gain due to your grazing management. Our raised yearlings do circles around sale barn purchased calves in performance on pasture.

The Commodity Side of Raising Yearlings

They have never been co-mingled with strange cattle, so no outside diseases are brought into the herd. Their rumen does not have to get adjusted to pasture, because they have not been on a feeder eating grain. All they know is a grass diet. They just perform better on pasture than what you can buy.

When you compare raised yearling calves' daily income versus custom grazed cows, the numbers really get exciting. If we take an average daily gain of 1.2 lbs per day for one year and compare that to running a custom grazed 1200 pound cow at .70 cents a day for one year, the difference in income is huge.

First of all you can run 1.5 steers on what one 1200 pound cow will eat. To keep the math simple let's assume you had 100 custom grazed cows that you were grazing for one year. The total custom grazed income for those 100 cows would be $25,500.

If we took 150 four-hundred-pound yearlings at 1.2 lbs gain per day for one year and received $1.10 a pound for them at market the total income would be $72,270. That is $46,770 more income from your grass operation because you grazed home raised yearlings.

Now, let me throw another bit of gas on the fire.

What if we used High Density Grazing and over time doubled our number of yearlings because each year we grew more grass as a result of HDG? This is where Ian Mitchell-Innes is right now.

If you don't want to direct market or raise grass-finished beef there is still a good living to be made by concentrating on High Density Grazing. Stay focused and build your mob, whether it is with owned livestock or a custom grazed mob. A golden opportunity is there.

Chapter 44
My Better Half

My wife, Jan, is a huge addition to our grazing enterprise. We were married in 2001 and we make a good team. She is a wonderful, smart partner, and is a great person to bounce ideas off of. She comes up with a lot of great ideas on her own as well.

Sometimes I will ask her a question about something and she has a totally different idea or direction to add to it. There have been lots of times when she has saved us money, time and work. Sometimes I will test her with an outlandish idea just to see what her thoughts are. She will almost always make my idea better or more simplified.

I may have a list of things written down that need to be done. Jan will always do her best to help accomplish these tasks in the order of importance.

She is terrific at helping work cattle. I would rather sort and load cattle with Jan, than any other person I know. She is extremely calm, moves slowly around the cattle without making any noise. When I get upset because a cow is not doing what I want it to do, she always smiles and reminds me to calm down.

She can string up a fence or take one down in the time it takes to tell you about it. I never saw a woman any stouter for her size than she is. She can handle her side of driving big corner posts right along beside me.

I try and never take her for granted, because I realize how lucky I am to have found her. She can work hard with me all day and still cook up a meal in 15 minutes that is to die for! I don't know how she does it. I can look in the refrigerator and not have a clue what to fix.

Jan amazes me with the laundry. She never has a big laundry day. When a full load accumulates, she washes it and hangs it out on the clothesline to dry. I've seen her out hanging up laundry on some pretty darn cold, sunny days. Back when I was single, laundry day was a big deal. Every Sunday the whole afternoon was blown because I had built up the laundry all week. I never hung up the clothes. They went right in the dryer. When we got married, I noticed our electric bill dropped about $25 a month. I wondered what caused that? Clothes hung outside to dry smell so much better and feel better against your skin than clothes dried in a dryer.

My diet has significantly improved since we got married as well. When I was single my freezer was filled with microwave dinners. They were fast and easy to prepare when I came in late at night from working on the farms with my coon light. Jan truly is a wonderful cook, and I appreciate every good meal that she prepares. Going from microwave meals to a prepared meal has been a real treat for me.

Our new grass-finished meat business is one of the areas where Jan excels. She is much more emotional and open with customers about their purchases. Women are better at marketing, I think. She sets realistic goals from one year to the next in ramping up our direct marketing of meat products. She started a website - greenpasturesfarm.net - as well.

I kid her a lot because she is an addicted dumpster diver. She will bring home things that are perfectly good, some items brand new - new jeans, shoes, shirts, along with numerous other goodies. She has packed home several new stock tanks that students at the university used for icing down brew. She has no pride when it comes to diving in a dumpster if she sees something of interest. I am so glad that Jan is a spendthrift, rather than a spender. Equity sure adds up faster when both of you are savers rather then spenders.

If I am going to be gone for several days, I set up numerous temporary paddocks on the various herds before I leave home. Jan has no trouble moving the herds while I am gone.

We had 200 head of dry cows that we were grazing for a fellow up north several years ago. He called and set up a date to haul them. He wanted just the cow numbers 1 through 100 sorted to haul. The rest of the cows were not to calve for another 60 days.

The trucks were to arrive at 5 p.m.. Jan and I went over that morning, and sorted out the cows. Some were pretty rank. Some would literally try and run over you. Jan never wavered. We had them sorted in about an hour with less than adequate handling facilities.

About 4:30 p.m. we drove over to prepare for loading the sorted cows. We were horrified to find the sorted cows had busted through a gate, and the whole herd was back together again peacefully grazing below the barn on about 40 acres. By this time it was 4:45, and the truckers would be there in 15 minutes. Truckers get paid to haul cows, not sit and watch you sort cows.

We quickly built a V out of temporary hot wire leading into the corral opening. Jan and I got a quarter-mile strand of white poly tape strung out between us and started gathering the cattle toward the V. The cows had just been caught and handled that morning so they knew what we were up to.

Jan and I walked the 200 cows back into the corral and started sorting again. The cows were irritated pretty good by then. Thirty minutes later we were sorting the last group of cows when we heard the semi-trucks coming down the driveway. I would not have gotten those cows sorted that quickly and efficiently without Jan.

On the really rank range cows that we calved several years ago, Jan reached her limit. We were scooting along about 15 mph beside a new calf with the ATV, and I told Jan to jump off and grab the calf. She looked at me like I had lost my mind. The old snorting cow was only about 30 feet behind us. It seemed like a reasonable request to me!

Jan has been soaked from head to toe, with mud from one end to the other, in freezing cold weather, you name it and she still performs very well. When I go speak at various grazing conferences around the country, sometimes Jan has to stay home and tend to the farms. No matter what comes up when I'm gone, she handles it in stride.

One of Jan's favorite tasks on the farms is to go out with her tree loppers and cut cedars that have come up out in the pastures. The birds eat the cedar seeds and spread them on the various pastures. Jan will round up her heeler dogs, grab her loppers, and can spend hours cutting cedars. The heelers love going with her to keep her company and chase bunnies. We don't have many baby cedars that make it to their second birthday with Jan patrolling the pastures.

She does have two weak spots. Don't let her get hungry while you are working. Her personality changes from pleasant to agitated very quickly. Once I get her fed, she is fine again.

Her other weak spot is caffeine. She must have a soda by 10 a.m. or she gets pretty ornery. I now make sure when we go out to work for the day that we have plenty of snacks and a soda or two. Keep her fed and supply her with a soda, and she is a happy camper. She's not a real high maintenance woman to say the least. I love my little dumpster diver very much.

Thank you, Jan, for putting up with me!

Index

Attitude 19-23
ATV reel holder 174-178
Bale unroller 35-38
Bulls 128, 223, 251-254, 260, 265
Burning 49-52
Calving 45, 109, 208-212, 256-257, 264
Commodity marketing 264, 270-271
Custom grazing 12, 14, 24, 53-55, 64, 75, 94, 223, 243, 245, 256, 258
Dog feeder 144-145
Dogs See Guardian dogs
Drought 44-48, 189
Dung beetles 192-195
Fencing and gear 22,28-29, 50, 56-63, 75, 98-104, 126-127, 164-178, 213, 224-230
Gamagrass 50, 204-207
Gates 45-47, 95-104, 164-178, 202
Genetics 14-15, 112-115, 246-250, 251-254, 258-262
Goats 14, 98-104, 120-125, 159
Grass-finished meat 264-269
Guardian dogs 92-93, 118, 128, 135-152, 209
Hay 19, 32, 35-43, 77
Herding - See Trailing animals
High Density Grazing 13-15, 49, 153-240, 271
Holistic Management 155-158, 231
Horses 64-73, 129-130
Leases 21, 24-30, 32, 74-82
Marbling 262-263, 267
Mitchell-Innes, Ian 13, 20, 155-158, 188, 219-223, 224-271
Mob grazing - See High Density Grazing
Multi-species 12-13, 91-97, 98-104
Parasites 12-13, 93, 96-97,105-108, 112-115
Pigs 13-14, 126-134, 221
Sheep 12-14, 91-97, 98-104, 105-111, 112-115, 116-199, 159
Trailing animals 224-230
Water 29, 166-168, 170, 172, 179-182, 191-192, 221

Questions
about grazing ??????
Answers *Free!*

While supplies last, you can receive a Sample issue designed to answer many of your questions. Topics include:

* Joel Salatin's Meadow Talk
* Flexible Grazing Cells
* Fly Management Without Pesticides
* Rejuvenating Depleted Soil
* Here Come the Dung Beetles
* Meat Goating
*Sheep Grazier
* Adaptable Gear for Pigs
* Happy Cows Don't Bawl
* And more

Green Park Press books and the *Stockman Grass Farmer* magazine are devoted solely to the art and science of turning pastureland into profits through the use of animals as nature's harvesters. To order a free sample copy of the magazine or to purchase other **Green Park Press** titles:

P.O. Box 2300, Ridgeland, MS 39158-2300
1-800-748-9808/601-853-1861
Visit our website at: www.stockmangrassfarmer.com
E-mail: sgfsample@aol.com

More from Green Park Press

COMEBACK FARMS, Rejuvenating soils, pastures and profits with livestock grazing management by Greg Judy. Grazing on leased land with cattle, sheep, goats, and pigs. High Density Grazing, fencing systems, grass-genetic cattle, parasite-resistant sheep. 280 pages. **$29.00***

CREATING A FAMILY BUSINESS, From contemplation to maturity by Allan Nation. Written with small, family businesses in mind. How to work with your spouse and children, emplyees and partners. Pre-start-up, pricing, production, finance and marketing. For anyone who wants to own their own business. 280 pages. **$35.00***

DROUGHT, Managing for it, surviving, & profiting from it by Anibal Pordomingo. Forages and strategies to minimize and survive and profit from drought. 74 pages. **$18.00***

GRASSFED TO FINISH, A production guide to Gourmet Grass-finished Beef by Allan Nation. How to create a year-around forage chain of grasses and legumes to create tender, flavorful grassfed products all year long virtually everywhere in North America. 304 pages. **$33.00***

KEEPING IT GREEN, A handbook for creating and managing irrigated pasture by Jim Gerrish. Covers all types of irrigation with pros and cons of each. Includes economics to determine best value and fit for your livestock class. 90 pages **$20.00***

KICK THE HAY HABIT, A practical guide to year-around grazing by Jim Gerrish. How to eliminate the most costly expense in operations anywhere in North America. Gerrish shares his experience gained in Missouri and Idaho. 224 pages. **$27.00*** Audio version: 6 CDs with charts & figures. **$43.00**

KNOWLEDGE RICH RANCHING by Allan Nation. Reveals secrets of high profit grass farms and ranches. Explains family and business structures. Nation shares knowledge he gathered over 30 years from successful financial ranchers worldwide. Anyone who has profit as their goal will benefit from this book. 336 pages. **$32.00***

LAND, LIVESTOCK & LIFE, A grazier's guide to finance by Allan Nation. How to separate land from a livestock business, make money on leased land by custom grazing, and how to create a quality lifestyle on the farm. Covers land-based financial issues to protect yourself from falling real estate prices without selling your farm or ranch. 224 pages. **$25.00***

MANAGEMENT-INTENSIVE GRAZING, The Grassroots of Grass Farming by Jim Gerrish. Takes graziers step by step through the MiG system. Using vivid images and detailed explanations, Gerrish begins with the soil, and advances through the management of pastures and animals. MiG basics: the power of stock density, extending the grazing season with annual forages. Chapter summaries include tips for putting each lesson to work. 320 pages. **$31.00***

MARKETING GRASSFED PRODUCTS PROFITABLY by Carolyn Nation. From farmers' markets to farm stores and beyond. Pricing, marketing plans, buyers' clubs, tips for working with men and women customers, and how to capitalize on public relations without investing in advertising. 368 pages. **$28.50***

NO RISK RANCHING, Custom Grazing on Leased Land by Greg Judy. Based on first-hand experience, Judy explains how by custom grazing on leased land he was able to pay for his entire farm and home loan within three years. How to find idle land to lease, calculate the cost of the lease, develop good water and fencing on leased land. Includes contract examples. 240 pages. **$28.00***

PADDOCK SHIFT, Revised Edition Drawn from Al's Obs, Changing Views on Grassland Farming by Allan Nation. A collection of timeless Al's Obs. 176 pages. **$20.00***

PA$TURE PROFIT$ WITH STOCKER CATTLE by Allan Nation. Profiles Gordon Hazard, who stocked a 3000-acre grass farm solely from retained stocker profits and no bank leverage. Economic theories are backed by real life budgets, including one showing how to double your money in a year by investing in stocker cattle. 224 pages **$24.95*** or Abridged audio 6 CDs. **$40.00**

QUALTIY PASTURE, How to create it, manage it, and profit from it by Allan Nation, Revised and updated by Jim Gerrish. Offers low-cost tactics to create high energy pasture to reduce or eliminate expensive inputs or purchased feeds. How to match pasture quality to livestock class, stocking rates for seasonal dairying, beef production, and multi-species grazing. Includes how to create a drought management plan. 300 pages. **$30.00***

THE CALENDAR OF THE YEAR-ROUND GRAZIER by Steven Kenyon. Although Kenyon only has a 4 month grazing season, he explains how to graze year-round in any region. Includes economic and financial tips for profitability. 80 pages. **$18.00***

THE MOVING FEAST, A cultural history of the heritage foods of Southeast Mississippi by Allan Nation. How using the organic techniques from 150 years ago for food crops, trees and livestock can be produced in the South today. 140 pages. **$20.00***

THE PRACTICAL SHEPHERD, Trials, errors and successes while maintaining profitability by Abram Bowerman. Pitfalls and challenges to become a profitable, successful shepherd. What to look for when buying sheep, a calendar for breeding, lambing and weaning, managing pastures, when to make a paddock shift and economics. 216 pages **$25.00***

THE USE OF STORED FORAGES WITH STOCKER AND GRASS-FINISHED CATTLE. by Anibal Pordomingo. Helps determine when and how to feed stored forages. 58 pages. **$18.00***

* All books softcover. Prices do not include shipping & handling.

Name _____

Address _____

City _____

State/Province_____Zip/Postal Code _____

Phone _____

Quantity	Title	Price Each	Sub Total
____	**Comeback Farms** (weight 1 lb)	**$29.00**	_____
____	**Creating a Family Business** (weight 1 lb)	**$35.00**	_____
____	**Drought (weight 1/2 lb)**	**$18.00**	_____
____	**Grassfed to Finish** (weight 1 lb)	**$33.00**	_____
____	**Keeping It Green** (weight 1 lb)	**$20.00**	_____
____	**Kick the Hay Habit** (weight 1 lb)	**$27.00**	_____
____	**Kick the Hay Habit Audio - 6 CDs**	**$43.00**	_____
____	**Knowledge Rich Ranching** (wt 1 ½ lb)	**$32.00**	_____
____	**Land, Livestock & Life** (weight 1 lb)	**$25.00**	_____
____	**Management-intensive Grazing** (wt 1 lb)	**$31.00**	_____
____	**Marketing Grassfed Products Profitably** (1½)	**$28.50**	_____
____	**No Risk Ranching** (weight 1 lb)	**$28.00**	_____
____	**Paddock Shift** (weight 1 lb)	**$20.00**	_____
____	**Pa$ture Profit$ with Stocker Cattle** (1 lb)	**$24.95**	_____
____	**Pa$ture Profit abridged Audio -- 6 CDs**	**$40.00**	_____
____	**Quality Pasture** (weight 1 lb)	**$30.00**	_____
____	**The Calendar of Year-round Grazing**	**$18.00**	_____
____	**The Moving Feast** (weight 1 lb)	**$20.00**	_____
____	**The Practical Shepherd** (weight 1 lb)	**$25.00**	_____
____	**The Use of Stored Forages** (weight 1/2 lb)	**$18.00**	_____
____	Free Sample Copy ***Stockman Grass Farmer*** magazine		_____

Sub Total _____

Shipping	Amount
1/2 lb	$3.40
1-2 lbs	$6.00
2-3 lbs	$7.00
3-4 lbs	$8.00
4-5 lbs	$9.60
5-6 lbs	$11.50
6-8 lbs	$15.25
8-10 lbs	$18.50

Mississippi residents add 7% Sales Tax _____

Postage & handling _____

TOTAL _____

Over 10 lbs or outside USA call for postage.

Please make checks payable to

Stockman Grass Farmer
PO Box 2300
Ridgeland, MS 39158-2300

1-800-748-9808
or 601-853-1861
FAX 601-853-8087